Analysis and Design of Gravity Flow Conduits and Buried Bridges

This book covers the structural analysis and design of buried gravity flow conduits, including traditional pipes, arches, box conduits, and buried bridges with spans up to 80 ft (25 m) and greater. The text primarily covers concrete, corrugated metal, and plastic conduits but is generally applicable to other materials. Applications include culverts, storm drains, sewers, and pedestrian and vehicular crossings.

The book is intended to introduce the subject to practitioners new to the field, as well as provide detailed information for those with prior experience. The opening chapter presents historical background and basic design models to introduce important concepts and then follows with chapters devoted to materials, soils, soil-conduit interaction, and guidance on the use of finite elements for analysis. Then design methods for evaluating soil-conduit systems are presented, along with guidance on important considerations during installation. The book concludes with field experiences of when things went wrong and why.

Analysis and Design of Gravity Flow Conduits and Buried Bridges offers a unified and comprehensive guide for practicing engineers working on buried pipe design, private consultants, and product manufacturers, as well as researchers in the area.

Timothy J. McGrath, spent over 40 years with Simpson Gumpertz & Heger Inc. before becoming Principal with TJMcGrath, LLC. Most of his career has been involved in research, design, and investigation of gravity flow conduits. Much of his work has been incorporated into the American Association of State Highway and Transportation Officials (AASHTO) LRFD Bridge Design Specifications and other design standards.

Analysis and Design of Gravity Flow Conduits and Buried Bridges

Timothy J. McGrath

CRC Press is an imprint of the
Taylor & Francis Group, an **informa** business

Cover image: Timothy J. McGrath

First edition published 2025
by CRC Press
2385 NW Executive Center Drive, Suite 320, Boca Raton FL 33431

and by CRC Press
4 Park Square, Milton Park, Abingdon, Oxon, OX14 4RN

CRC Press is an imprint of Taylor & Francis Group, LLC

ISBN: 978-1-498-74782-0 (hbk)
ISBN: 978-1-032-93554-6 (pbk)
ISBN: 978-0-429-16261-9 (ebk)

DOI: 10.1201/9780429162619

Typeset in Sabon
by SPi Technologies India Pvt Ltd (Straive)

Contents

Preface

The world's infrastructure includes millions and perhaps billions of pipes, culverts, and buried bridges. I did not set out to work in this area, and I suspect that is true for most people and perhaps everyone working in this field. The subject is not taught in any detail at the undergraduate university level, and there is no standard text to serve as an introduction to the field. In other words, most of us were given an assignment related to these structures, and after laboring to collect pertinent information from reports, papers, and codes, we wound up with a new area of expertise – which is how I started. While a circular cylinder seems like a simple structure, its behavior is more complicated when it is embedded in soil. It has held my interest for many years.

This book began as a full-day seminar on buried pipe design to address the need for entry and advanced-level guidance on the design of gravity flow conduits. This book expands the content of the seminar material to serve as a detailed primer on the subject but also provides more advanced details that experienced users should find helpful.

Names and Shapes – As explained in Chapter 1, "conduit" encompasses all the shapes, sizes, and types of pipes, culverts, and buried bridges that are discussed in this book. It seems awkward to use but hopefully will be a constant reminder of the breadth of the field. "Soil-conduit interaction" is used in place of the more common phrase soil-structure interaction to emphasize the combined strength of conduits and the surrounding soil – i.e., it is the combination of the conduit properties and soil support that provides a viable structure.

Most of the behaviors discussed herein are presented in terms of round conduits in part because the research that identified the behaviors was focused primarily on round culverts and also because it is often easier to grasp the behavior of a circle than other shapes.

State-of-the-Art – The American Association of State Highway and Transportation Officials (AASHTO) LRFD Bridge Design Specifications are used as a primary reference for conduit design procedures. It is one of the few codes that provides design procedures for concrete, metal, and plastic gravity flow conduits. The design procedures for these conduits evolved over the last century, generally being developed for specific products. During that evolution, design approaches have taken advantage of more sophisticated analytical tools as they became available. Metal conduits in particular show this growth where "long-span" conduits are designed prescriptively without load factors or calculations, metal box conduits are designed with simplified equations limited by shape and size, and "deep corrugated" metal conduits are analyzed and designed with finite element analysis. Concrete pipes have the oldest conduit design method, but one that is now supplemented by a newer "standard installation" approach. The old method is still in regular use because it is effective and can be used to analyze installations not covered in the newer method. For both metal and concrete, the effectiveness of the older methods has kept them in use even though more sophisticated approaches could be implemented.

The diversity of design methods does not just lie with the conduit. The names of soil groups used to identify backfill quality are different for each type of conduit. The field of conduit design would greatly benefit if the design methods could be made more consistent. Uniform soil groups would be a good place to start.

Acknowledgments – Many people and organizations have supported and contributed to this book. My greatest thanks go to Professor Ian Moore, my colleague and collaborator for many years, who contributed Chapter 6 on finite element analysis of conduits and also improved all of the other chapters through his thoughtful review, comments, and suggestions.

Dr. Mike Katona and I first spoke in the mid-1970s – his active mind, thoughtful comments and questions, and our collaborations together are strong positive memories. There are too many other individuals to list – colleagues, fellow employees at Simpson Gumpertz & Heger Inc. (SGH), members of industry, academics, and more – many of whom I consider friends. Comments, discussions, questions, encouragement, and disagreements with these people all contributed to my work in a positive way.

Three people share responsibility for the arc of my career: Dick Chambers, who introduced me to plastics; Dr. Frank Heger, who introduced me to concrete pipe while we collaborated on the development of reinforcing design equations; and Professor Ernie Selig, who opened my eyes to soil behavior and eventually became my PhD advisor. I could not have written this book without their contributions to my education and career.

Thanks go to the many organizations that have granted me access to use graphics from their own codes, manuals, reports, and papers. Alphabetically they include AASHTO, the American Concrete Institute (ACI), the American Concrete Pipe Association (ACPA), the American Society of Civil Engineers, (ASCE) ASTM International (ASTM), the American Water Works Association (AWWA), the US Federal Highway Administration (FHWA), Iowa State University, the National Cooperative Highway Research Program and Transportation Research Board (NCHRP/TRB), the National Corrugated

Steel Pipe Association (NCSPA), Relativity Publishing, and the Uni-Bell PVC Pipe Association (Uni-Bell).

My special thanks to Roberta Ferriani, who assisted with necessary administrative tasks, and to SGH, where I spent 43 years mostly working on pipe and culvert issues, who gave me access to their library and, in particular, to Joan Cunningham, SGH's librarian who seemed to be able to instantly produce any paper or report that I requested.

Chapter 1

Introduction

1.1 OVERVIEW

Gravity flow conduits have been with us for many centuries. Well before the Christian era, societies constructed conduits and aqueducts to bring water to cities and to dispose of sewage. Copper water conduits were used in India between 3,000 and 4,000 BCE, and the earliest known clay conduits were used in Babylonia. Perhaps the best-known ancient conduits are the Roman aqueducts and sewers. The need to deliver drinking water from source to consumer or wastewater from consumer to disposal or now, fortunately, treatment, was required as people congregated in cities. Today, drinking water is almost always delivered in pressure conduits, but stormwater and wastewater still operate primarily through gravity flow conduits. Gravity flow culverts carry streams and stormwater under roadways. Agriculture uses underdrains to control water in fields. In the early 1900s, researchers began to understand that the soil in which they were embedded could assist in resisting earth and vehicle loads.

The key to understanding the behavior of buried conduits is recognizing that the soil around the conduit both imposes load on the conduit and provides support to resist deformation and cracking. In the parlance of load and resistance factor design, the soil is both load and resistance. The early researchers saw this role for soil and began the process of using it to design economical installations. The study of this behavior is typically called soil-structure interaction, but that is misleading, as the "structure" is only viable if the soil provides support to the conduit. In this text, the structure is called a soil-conduit system as a reminder of the importance of both elements to long-term performance. If designed and constructed properly, a soil-conduit system can support much more load than a conduit alone. This book covers the behavior and design of "conduits" – a term used here to include the various sizes, shapes, and structural systems that are presented. Some of these structures have spans up to 80 ft (24 m) or more, earning the title "buried bridges," but these large structures depend on soil support much like smaller conduits. Either through load sharing or in their dependence on lateral soil support, buried bridges are soil-conduit systems.

DOI: 10.1201/9780429162619-1

1.2 CODES AND UNITS

Many standards writing organizations, trade associations, and individuals have developed design methods for conduits. It is not possible or productive to present or evaluate them all. The *AASHTO LRFD Bridge Design Specifications* (AASHTO, 2020, called AASHTO LRFD throughout) include design methods for metal, concrete, thermoplastic, and fiberglass (GRP) conduits. While AASHTO LRFD is intended for the design of bridges and culverts, the culvert design methods presented are widely applicable to most gravity flow conduits; thus, this standard will be the primary reference for the design methods presented throughout this monograph. AASHTO LRFD references *AWWA Manual of Water Supply Practice M45 Fiberglass Pipe Design* (AWWA, 2020) for details of fiberglass conduit design and some material presented will reference that manual. The American Concrete Pipe Association does not write standards but has done an excellent job of documenting the historic and current methods of concrete pipe design in the *Concrete Pipe Handbook* (1998). Other organizations, trade associations, and individuals have developed design methods for gravity flow conduits. No opinion is offered on these methods, and leaving them out of the text is not a judgment on their suitability for conduit design.

The text uses US customary units, also known as in-lb units, as primary and includes SI units parenthetically. A number of equations are not dimensionally consistent. Some equations were developed with constants that carry unstated units, and properties of concrete are often based on the square root of the compressive strength, which still carries the units of force per unit area. These equations have been provided with additional constants to make the equation accurate in either system of units if the listed variable units are used in calculations.

1.3 ORGANIZATION

This book is organized to introduce conduits and soil as separate subjects before exploring the interaction of the two and, finally, their design and installation. Thus, it begins with the basics of soil-conduit interaction, conduit materials, and soil, each as a separate subject in Chapters 2, 3, and 4, followed by chapters with more detail on analysis, design, and construction. Finally, a few experiences are presented that demonstrate the importance of paying attention to the lessons learned by others.

Chapter 2 serves both to provide a basic introduction to the key issues discussed in more detail in the rest of the book and to give credit to those early researchers who investigated that behavior by developing new testing and measurement techniques to gather data on soil-conduit interaction. With the insights gained from their tests, they used that data to develop analytical tools that engineers use to this day for conduit design. Presenting the early research introduces the primary concepts of conduit behavior and soil-conduit interaction. This chapter is a good place for readers new to the field to get started and will seem familiar to those who have worked in the field.

Chapter 3 presents concrete, steel, thermoplastic, and fiberglass (GRP) as the materials used to manufacture conduits. The basic material stiffness, strength, and failure modes are presented and then how those material properties create products with different behaviors when formed into a ring. A ring, the shape of the vast majority of conduits, often serves as a tool in understanding behavior, but the insights taken from that behavior are applicable to all the other shapes of conduits.

Chapter 4 begins by introducing systems used to classify soils in groups based on expected behavior. Particle size for sands and gravels and plasticity for silts and clays are the key determinants in predicting behavior in conduit installations, but the density of those particles after placement and compaction is equally important. Laboratory compaction tests serve to define how much energy is required to achieve density and predict the density that can be achieved in the field. Once the soil classification is set and the possible densities are known, the stiffness of a soil can be evaluated. This is accomplished with simple linear constitutive models or more complex nonlinear models that are suitable for computer analysis of behavior.

After the pipe and soil stiffnesses are characterized separately, their contributions to soil-conduit interaction are explored in Chapter 5. A simple elastic model of a ring in an infinite medium demonstrates how this interaction can be characterized and where concrete, metal, and plastic conduits fit into a single continuum of behavior. This continues with presentations of the various models that have been created for use in designing soil-conduit systems. Chapter 6 explores the modeling of soil-conduit systems through finite element analysis. This tool allows engineers to analyze the multitude of combinations of conduits of various sizes, shapes, and stiffness when embedded in soil masses that can be defined in great detail, including in situ and backfill soils, zones of varying density and stiffness, local variations that might include voids or rocks, and any other zone that an engineer wishes to model.

Chapter 7 brings all the information from the prior chapters together to provide guidance on designing a soil-conduit system, that is, how to combine soils with conduits to provide a stable long-term installation. At this stage, an engineer evaluates the in situ soil conditions, selects backfill from locally available materials, and specifies compaction requirements. A conduit material is selected and evaluated to ensure it will not reach a state of failure under the as-built conditions.

Chapter 8 discusses the installation process. Installing conduits can be complex and must address the many underground conditions that might be encountered during construction. The focus of this chapter is on how the installation process can affect the behavior of the soil-conduit system. The techniques to address these conditions are not addressed in detail, as there are many manuals, standards, and books available that cover these issues in great detail. The important goal of installation is to achieve the system that was designed. Monitoring and inspection are important steps toward achieving this.

So much can be learned from studying what has happened before. George Santayana stated it so well with his famous quote: "Those who cannot remember the past are condemned to repeat it." This quote speaks both to the importance of passing high school history class to avoid taking it again, as well as understanding global, social, and political matters. In the present case, owners, engineers, and contractors can learn the importance of following good design and construction practices to achieve successful soil-conduit systems by reviewing instances where things went wrong. Chapter 9 presents a few such instances and explains what happened to cause the problems. The goal of this final chapter is to drive home the importance of understanding soil-conduit systems and the processes by which we design and build them so they will be built to successfully serve their intended use.

REFERENCES

AASHTO (2020) *LRFD Bridge Design Specifications*, 9th Edition, American Association of Highway and Transportation Officials, Washington, DC.

ACPA (1998) *Concrete Pipe Handbook*, American Concrete Pipe Association, Irving, TX.

AWWA (2020) *Fiberglass Pipe Design, AWWA Manual of Water Supply Practices M45*, 3rd Edition, American Water Works Association, Denver CO.

Chapter 2

Basics of soil-conduit systems

The conduit industries of primary interest in this chapter began to develop in the 1800s – first with concrete and clay, and late in the century with corrugated steel. Plastic conduits were developed mostly in the second half of the 1900s. Early in the 1900s, engineers recognized and began to investigate the role that soil embedment played in the performance of gravity flow conduits and developed new tests to evaluate design concepts. These engineers developed the first known models of soil-conduit interaction and developed design methods still in use today. It is well worth using this early research both to pay tribute to the engineers who did the work and to introduce the important concepts of buried conduit design. While many conduit designs are now completed using finite element models, an understanding of the concepts of soil-conduit interaction, as developed in early research, is a significant aid in properly interpreting computerized analysis and test results.

This Chapter presents the key features of soil-conduit interaction in terms of circular conduits, as that was the focus of research at the time of development. There are now many conduit shapes and sizes. Conduit spans can be greater than 70 ft (21 m). Shapes include rectangular three- and four-sided concrete box sections, concrete arches, metal long-span structures of many shapes with open and closed bottoms and even thermoplastic arches that are used for rainwater detention. These structures are introduced and discussed at appropriate locations in the following Chapters.

2.1 DEFINITIONS

Important terms used throughout this book are defined here. Terminology for round conduits is presented in Figure 2.1 and is generally applicable to other conduit shapes. Terminology for trench geometry is presented in Figure 2.2.

Figure 2.2a is the primary reference for flexible conduit trench terminology. Figure 2.2b shows the terminology for concrete conduits in embankment installations as used in US practice.

DOI: 10.1201/9780429162619-2

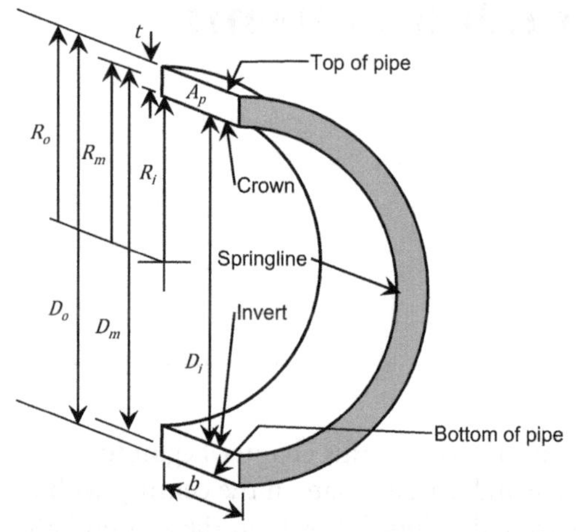

t = conduit wall thickness, in., mm
D_i, D_m, D_o = inside, mean, and outside diameter, in., mm
R_i, R_m, R_o = inside, mean, and outside radius, in., mm
A_p = conduit wall cross-sectional area per unit length, in.^{2}/in., mm^2/mm
I_p = conduit wall moment of inertia per unit length, in.^{4}/in., mm^4/mm
b = unit length, 1 in., 1 mm

A_p = t for solid wall conduit
$Ip = t^3/12b$ for solid wall conduit

A_p and I_p are calculated or taken from product tables for profile wall conduit.

For profile wall conduit, R_m and D_m are measured to the centroidal axis of the conduit wall section.

Figure 2.1 Terminology for conduits.

In situ soil – The in situ soil is the soil mass in which a trench is excavated for conduit installation or on which an embankment is constructed. In situ soils typically have been in place for a long period of time, allowing settlement and consolidation to take place. With some exceptions (e.g., organic soils, soft clays), in situ soil strength and stiffness are generally greater than that of freshly placed soils. If suitable, conduits may be laid directly on the in situ soil.

Foundation – If the in situ soils are soft or unstable, they may be removed and replaced with suitable soil. Imported foundation soils are often not required. If required, the foundation soil can also be used as the bedding material if the gradation is compatible with the type of conduit to be installed.

Embedment zone – The embedment zone encompasses all the soil in contact with the conduit. For concrete conduits, there are minimal restrictions on the backfill above the springline.

Bedding – The bedding soil is the layer on which the conduit is placed. The gradation must be suitable for contact with the conduit.

Haunch zone – The haunch zone for flexible conduits (Figure 2.2a) is the area directly below the conduit and above the bedding. A manual effort is often required to place and compact backfill in this zone. For concrete conduits (Figure 2.2b), the haunch zone extends to the outer edge of the embedment zone (Figure 2.2b) for embankment installations or to the trench wall for trench installations.

Sidefill – The sidefill is the soil placed at the sides of the conduit. This zone and the haunch zone provide critical lateral support to flexible conduits.

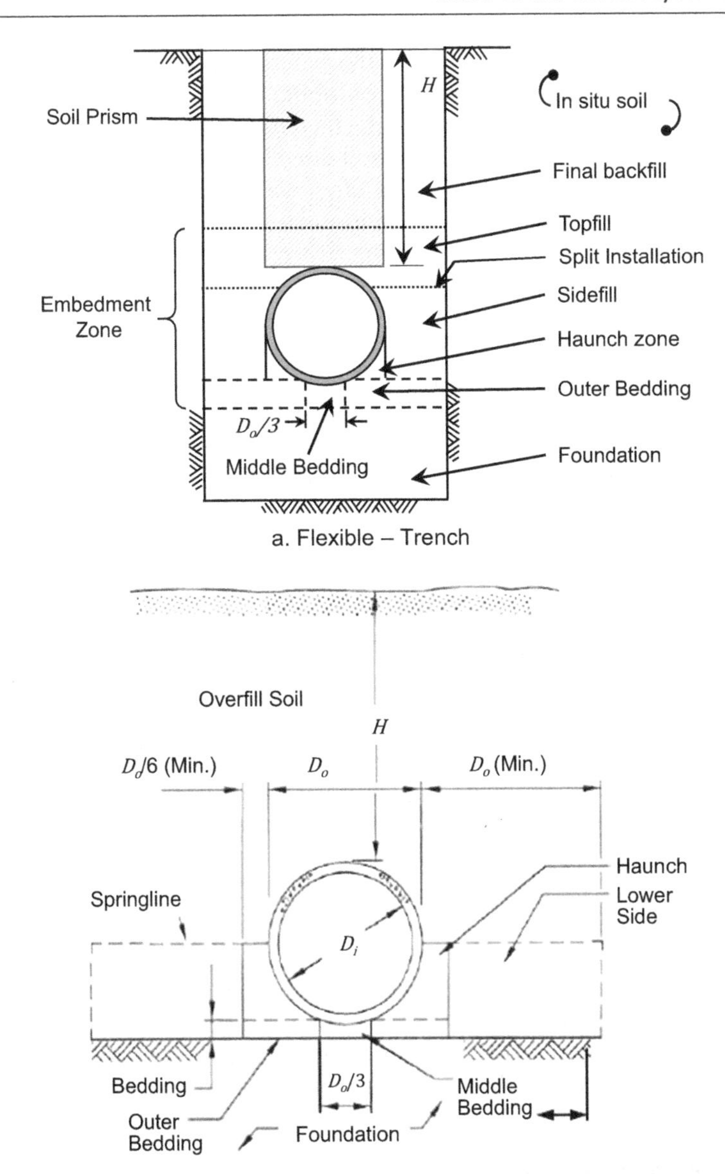

Figure 2.2 Terminology for (a) trench and (b) embankment installations.

Top fill – The top fill is the first backfill placed over the top of the conduit. The top fill is most often the same material as the sidefill. The top fill provides protection from the final backfill, which may not be suitable for contact with the conduit. Concrete conduits often have only limited restrictions on the backfill above the springline.

Final backfill – The final backfill is the material placed above the embedment zone. This material does not contribute to conduit support. Gradation and compaction are based on the final use of the ground surface.

Soil Prism – The soil prism is the soil directly over the conduit. The weight of the soil in the soil prism is often used as a reference point for the load on a conduit in an actual installation:

$$W_{sp} = D_o \gamma_s H, \tag{2.1}$$

where

W_{sp} = Weight of the prism of soil directly above the top of conduit, lb/ft, kN/m (Note that at times load is computed as lb/in., kN/mm – always make sure units are consistent)

D_o = Outside diameter of conduit, ft, m

γ_s = Unit weight of soil, lb/ft^3, kN/m^3

H = Depth of backfill above top of conduit, ft, m

For large structures, the soil load in the area below the crown and above the springline can be significant, in which case, for circular conduits,

$$W_{sp-t} = D_o \gamma_s \left(H + 0.11 D_o\right), \tag{2.2}$$

where

W_{sp-t} = Weight of all soil above the conduit springline, lb/ft, kN/m

The earth load is often non-dimensionalized against the weight of the soil prism by calculating the vertical arching factor:

$$VAF = \frac{W_E}{W_{sp}}, \tag{2.3}$$

where

VAF = vertical arching factor – ratio of vertical earth load carried by conduit to the soil prism load

W_E = vertical earth load carried by conduit, lb/ft, kN/m. W_E is determined as the total axial force (thrust) at the conduit springlines.

2.2 RIGID CONDUITS

Concrete and clay pipes are rigid conduits. The design of clay pipe for loads is similar to that of concrete, although clay pipe does not have reinforcement to limit deformation after cracking. This section focuses on concrete pipe. The *Vitrified Clay Pipe Engineering Manual* (*NCPI*, 2017) provides details on clay pipe.

Rigid conduits have high bending stiffness and are designed to resist deformation primarily through bending moments. The high bending stiffness results in somewhat concentrated soil pressures on the conduit, as documented in early research (Figure 2.3), which shows high vertical pressure at the top and

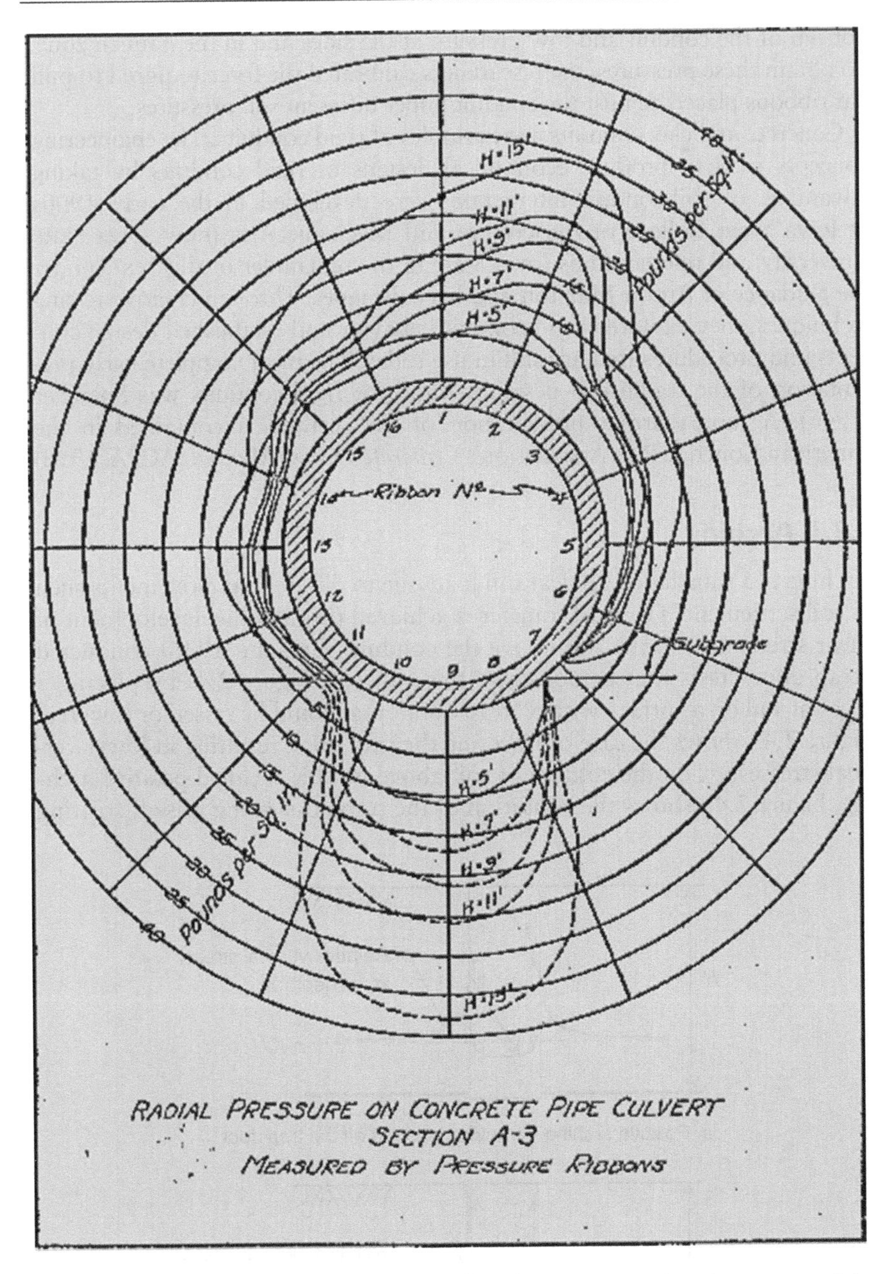

Figure 2.3 Early data on soil pressure distribution around concrete conduit. (Marston, 1930, with permission Iowa State Library Special Collections and University Archives)

bottom of the conduit and low pressure at the sides and in the haunch zone. To obtain these pressures, the researchers calibrated the force required to pull flat ribbons placed against the conduit under different soil pressures.

Concrete and clay conduits are examples of rigid conduits. The engineering concepts used to produce economical designs of rigid conduits by taking advantage of soil-conduit interaction were developed in the early 1900s at Iowa State College of Agriculture and Mechanic Arts (now Iowa State University and referenced as Iowa State in the remainder of this text) under the guidance of Anson Marston and his colleagues. Through creative testing techniques, they explored rigid conduit behavior and established design concepts and procedures that are still in use today. The most complete early presentation of the traditional design method for rigid conduits was Spangler (1946). A more current presentation of the method is contained in the American Concrete Pipe Association's *Concrete Design Manual* (ACPA, 1998)

2.2.1 Arching

Arching is a transfer of vertical soil load due to differential movements such as soil settlement. The load transfer is achieved through the development of shear stresses (τ) in the soil above the conduit. Terzaghi (1936) conducted "trap door" tests that demonstrate this concept (Figure 2.4) by placing a layer of soil on a surface with a "trap door" that could be raised or lowered. Figure 2.4a shows the case of lowering the trap door, resulting in lower vertical stresses, σ_v, on the column of soil above it. This is called positive arching. Figure 2.4b shows the condition of the trap door being raised, creating

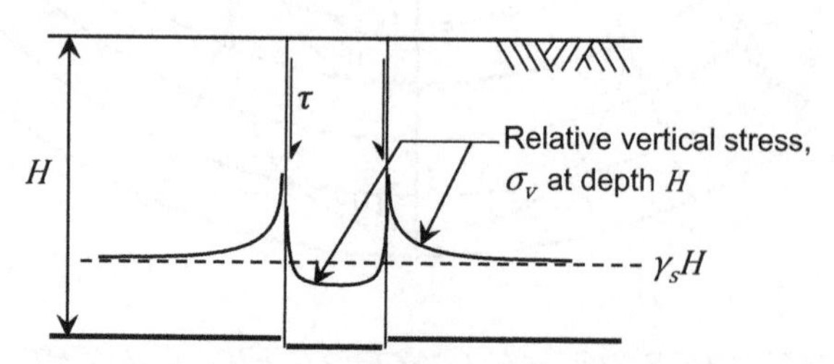

a. Positive arching – transferring load off the trap door

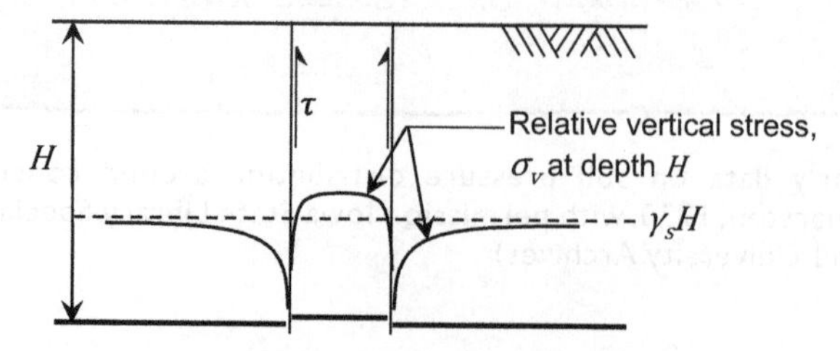

b. Negative arching – transferring load onto the trap door

Figure 2.4 Soil stress distribution in trap door tests.

an increased load on the door: negative arching. Load transfer off of, or on to, the trap door is achieved through the development of shear stress (τ) from the soil over the trapdoor onto the surrounding soil.

Terzaghi's trap door tests were conducted in 1936, but about 20 years earlier, Marston (Marston and Anderson, 1913) and his colleagues at Iowa State University had investigated how arching applied to buried conduit design and began the development of equations for soil load calculations.

Marston first considered the condition of a buried conduit in a trench installation. Backfill placed in a trench over a conduit will settle with time, resulting in shear stresses at the trench wall that transfer the vertical load into the undisturbed soil, just as in the Terzaghi positive arching test. One of the test configurations used to evaluate this condition is shown in Figure 2.5, where the test conduit was hung from support rods and weighed as

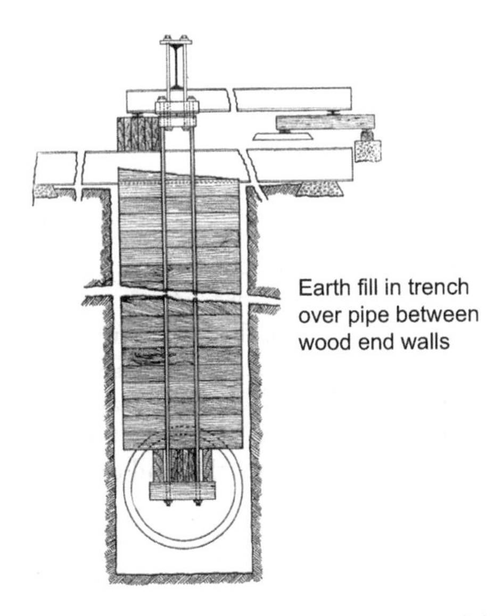

Figure 2.5 Marston trench load test apparatus. (Marston and Anderson, 1913, with permission Iowa State Library Special Collections and University Archives)

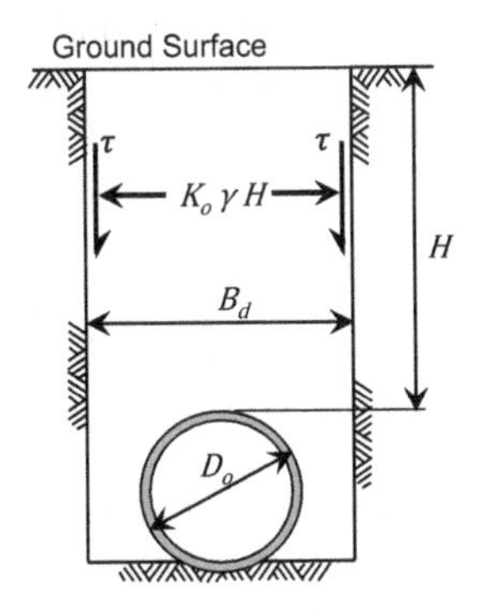

Figure 2.6 Trench installation.

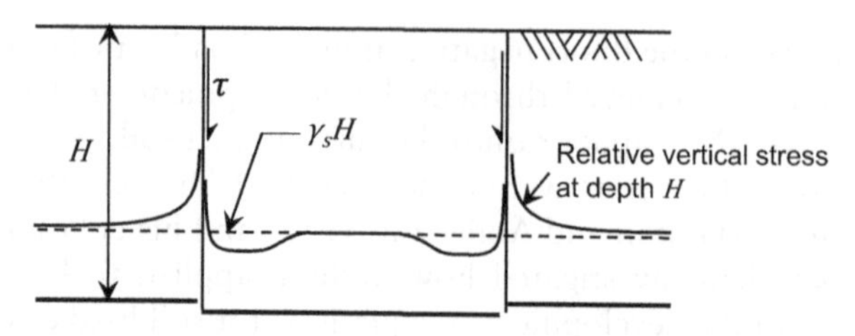

Figure 2.7 Wide trap door condition.

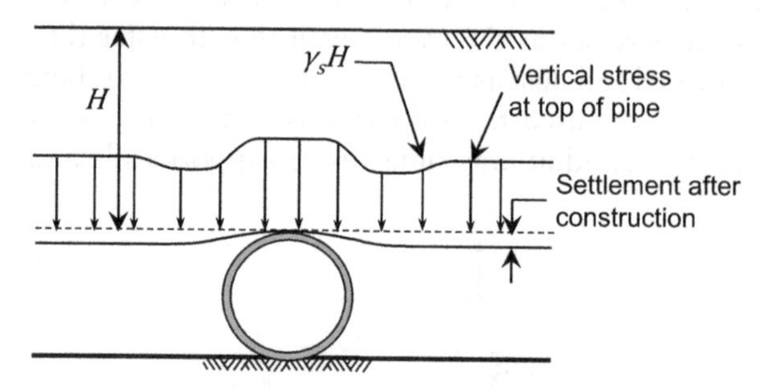

Figure 2.8 Embankment or wide trench condition.

backfill was added and consolidated. Schematically, Figure 2.5 can be drawn as a standard conduit installation (Figure 2.6).

In the case of a trench installation, the shear stresses of interest are those at the trench wall/in situ soil interface, as shown in Figure 2.6. This is the equivalent of the column of soil over Terzaghi's trap door. The shear stresses at the trench wall interface reduce the earth load on the conduit. In narrow trenches, these shear stresses can reduce the load on the conduit to less than the soil prism load (W_{sp}), i.e., $VAF < 1.0$.

The stress redistribution in a conduit trench or over a trap door is localized to a width near the edge of the trap door (or the trench wall). If the trap door of Terzaghi's experiments is wide (Figure 2.7), the vertical stress in the central region over the door returns to the free field stress ($\gamma_s H$).

When a rigid conduit is installed in an embankment condition (Figure 2.8), a plane through the soil at the top of the conduit during construction will undergo settlement, which again results in shear stresses that increase the load on the conduit. In the embankment condition, VAF is typically greater than 1.

2.2.2 Vertical soil load

Research at Iowa State established four installation types (Figure 2.9) that represent the major configurations considered for design. These include:

Trench – the trench installation is perhaps the most common type. Most trench installations qualify as narrow trenches, and the earth load on the conduit is less than the weight of the soil prism ($VAF < 1$).

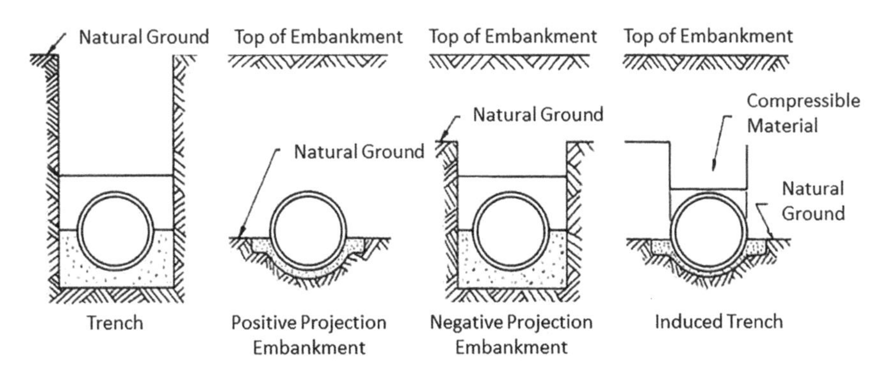

Figure 2.9 Installation conditions for buried conduits. (ACPA, 1998, with permission)

Positive projecting embankment – Conduits are often required under highway embankments, in which case a conduit might be laid on natural ground and the embankment placed beside and above it ($VAF > 1$). Conduits installed in a wide trench are designed as positive projecting installations.

Negative projecting and induced trench – Negative projecting and induced trench conditions are both installations that can increase positive arching relative to a standard trench or embankment, i.e., to further reduce load on the conduit. The negative projecting embankment installation consists of conduit installation in a narrow subtrench prior to placing an embankment. In this case, the earth load due to the embankment is supported by the in situ material at the sides of the conduit subtrench. If the backfill over the conduit but below the elevation of the original ground surface is left uncompacted, the reduction in load on the conduit is further reduced. The induced trench condition intentionally places a zone of compressible backfill material over the conduit, which allows for settlement and a subsequent reduction in load on the conduit. These installations are not widely used, as they require careful evaluation of the in situ soil, the compressible material, and the trench configuration to assure successful long-term performance. The relative compressibility of the two zones must be adequate to gain the desired behavior, and the load reduction must be stable over time. The negative projection and induced trench conditions are not used for flexible conduits.

2.2.3 Lateral soil support

In addition to evaluating vertical soil load, early researchers understood that lateral soil load provides support to a conduit and reduces the bending moments. In a narrow trench, where the earth load on the conduit is

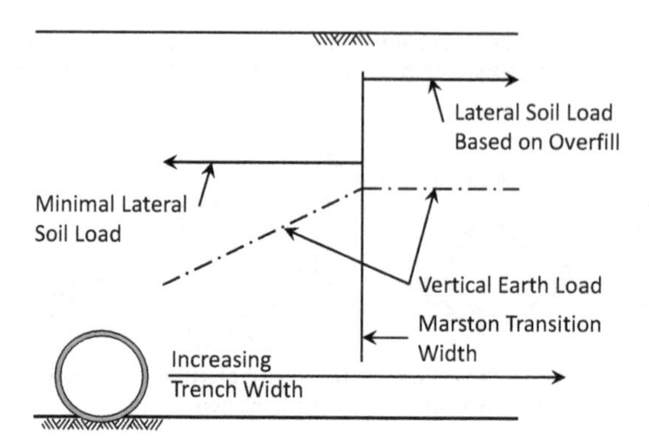

Figure 2.10 Change in load and lateral support with increasing trench width.

generally less than the embankment load, the lateral soil pressure on the conduit is minimal, as the vertical soil load is mostly carried by the conduit or is transferred to the trench wall. The lack of vertical stresses in the sidefill limits the lateral pressure applied to the conduit wall. In a wide trench (embankment) condition, lateral soil pressure was found to be more significant as the sidefill carries more of the vertical load, which increases the lateral pressure to the conduit. However, no transition was provided to gradually change the lateral pressure on the conduit as the trench widened from the narrow condition to the wide condition, resulting in a stepwise change in lateral soil load at the "transition width" when the vertical load calculation changed from trench to embankment (Figure 2.10). This approach resulted in designs near the transition width that could be quite conservative by underestimating the lateral pressure on the conduit and was later addressed by introducing a linear increase in lateral support as the trench widens (ACPA, 1998). This was accomplished by introducing a transition in the bedding factor, B_f, introduced in the following section, as the trench width increased from narrow to the transition width.

2.2.4 Bedding support

The width and quality of support provided to a buried conduit by the bedding and compaction of backfill in the haunch zone and the sidefill are critical factors for the design and good performance of rigid conduits.

Trench Condition – In the Marston method for trench design, a bedding factor, B_f, was used to address the bedding and lateral support condition. Common bedding conditions (Figure 2.11) were identified, and each condition was associated with a specific bedding factor. The bedding factor is the ratio of the invert bending moment for the in-ground condition to the invert bending moment in the three-edge bearing test, a standard quality control test introduced in Section 3.2.2. Thus, a

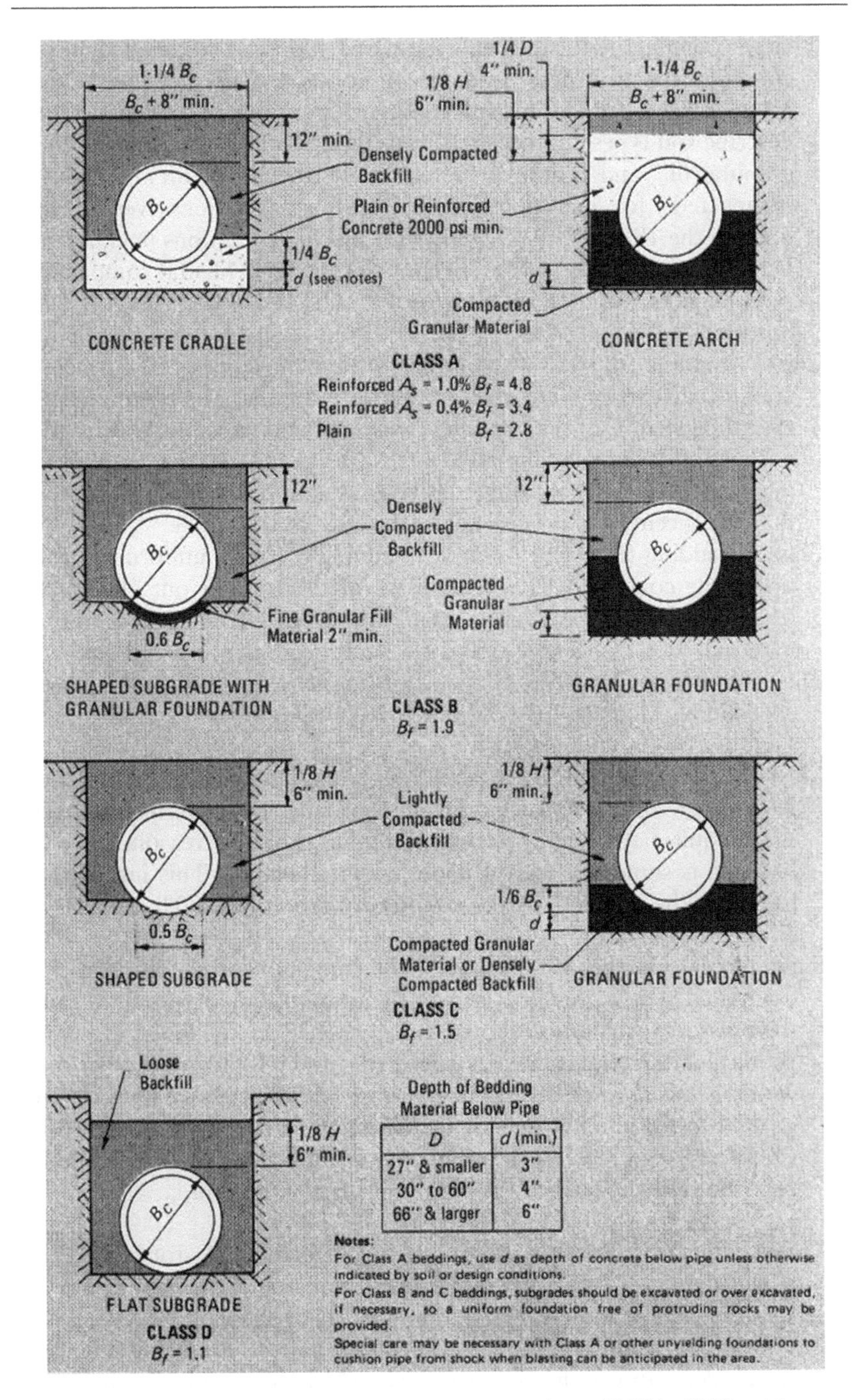

D	d (min.)
27" & smaller	3"
30" to 60"	4"
66" & larger	6"

Figure 2.11 Traditional beddings for concrete conduits. (ACPA, 1998, with permission)

larger bedding factor indicates improved support and lower required conduit strength. These bedding conditions and the bedding factors associated with them are still in use today; however, terms used to describe soil types and densities were limited and vague, such as "fine granular fill" and "granular material" for trench backfill and "lightly compacted" for density. Updated bedding conditions were developed in the 1980s (Heger, 1988) with specific gradations (e.g., ASTM D2487 and AASHTO M145, Section 4.1) for backfill and compaction requirements based on the Proctor test (ASTM D698, AASHTO T99, Section 4.2).

Class A bedding (B_f = 2.8–4.8) – Class A bedding consists of a concrete cradle that can provide uniform vertical support and restrain horizontal deflection of the conduit. Concrete bedding is not widely used because of the expense but also because design procedures are not well established. Further, in some designs, conduit cracking has occurred due to concentrated point loads at the edge of the cradle.

Class B bedding (B_f = 1.9) – Class B bedding provides uniform support under the conduit over at least 60% of the conduit outside diameter and lateral support using sidefill of "densely compacted granular material."

Class C bedding (B_f = 1.5) – Class C bedding relaxes the minimum width of bedding support to 0.5 times the conduit diameter and allows "lightly compacted backfill."

Class D bedding (B_f = 1.1) – Class D bedding places almost no bedding, backfill, or compaction requirements on the embedment material and thus requires the highest strength conduit. For this reason, Class D bedding is sometimes called impermissible bedding. This installation increases the demand on the conduit by more than 70% relative to Class B bedding.

Embankment Condition – As noted, the embankment condition uses a calculated bedding factor based on the bedding support condition and the pressure distribution shown in Figure 2.12. In the figure, K_o is the coefficient of lateral earth pressure at rest, and p is the projection ratio, which is the ratio of the projection of the conduit above undisturbed ground to the conduit outside diameter. These terms are discussed in Chapters 4 and 7. D_o is shown as the outside diameter of the conduit but would be the outside rise for a noncircular conduit.

While fading from current use, the range of bedding factors based on varying trench and embedment conditions demonstrates the importance of understanding soil-conduit interaction in the design phase.

The traditional beddings and associated calculations are being replaced by more modern approaches to materials and analysis and computerized calculations. However, understanding these traditional, simplified methods provides considerable insight into soil-conduit behavior. It is well worth the

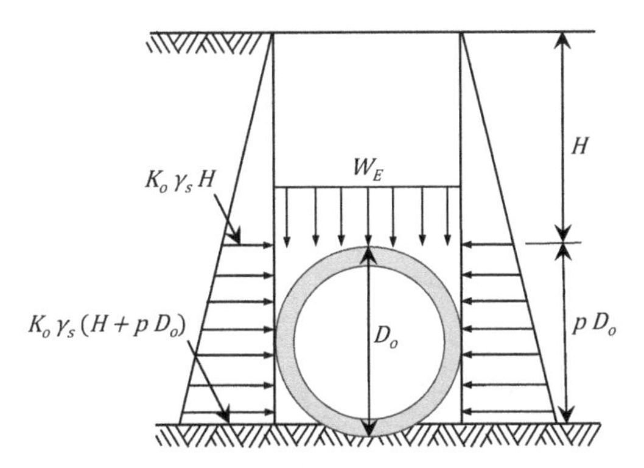

Figure 2.12 Marston pressure distribution for embankment conditions. (ACPA, 1998, with permission)

effort to study and understand them. Such understanding will aid the conduit designer in applying and interpreting the results of the newer approaches.

2.3 FLEXIBLE CONDUITS

Flexible conduits have low bending stiffness and deform under load, resulting in more uniform pressure distributions (Figure 2.13) relative to rigid conduit (Figure 2.3). Note that both types of conduit have a reduced pressure in the haunch zone. Much research has focused on methods to predict the deflection of flexible conduits under earth load as a design criterion and for field quality control.

2.3.1 Deflection

Corrugated steel conduits were the first flexible conduit mass-produced for roadway conduits and other larger-diameter applications. Early installations of these conduits showed the importance of controlling deflection to achieve good performance. An early observation noted that these conduits were likely to collapse if deflection (reduction in vertical diameter during and after backfilling) exceeded 20%. On the belief that 20% deflection was a stability limit, a somewhat arbitrary safety factor of 4 was adopted, resulting in a service load deflection limit of 5%. This 5% limit on reduction in vertical diameter became established as an industry standard that was later adopted for plastic conduits due to its familiarity to designers. A single deflection limit for the flexible conduit is convenient but is not based on achieving a uniform factor of safety. The relationship between safety and deflection is discussed in Chapter 7 on design. Controlling and monitoring deflection in flexible conduits does ensure a level of soil support around the conduit

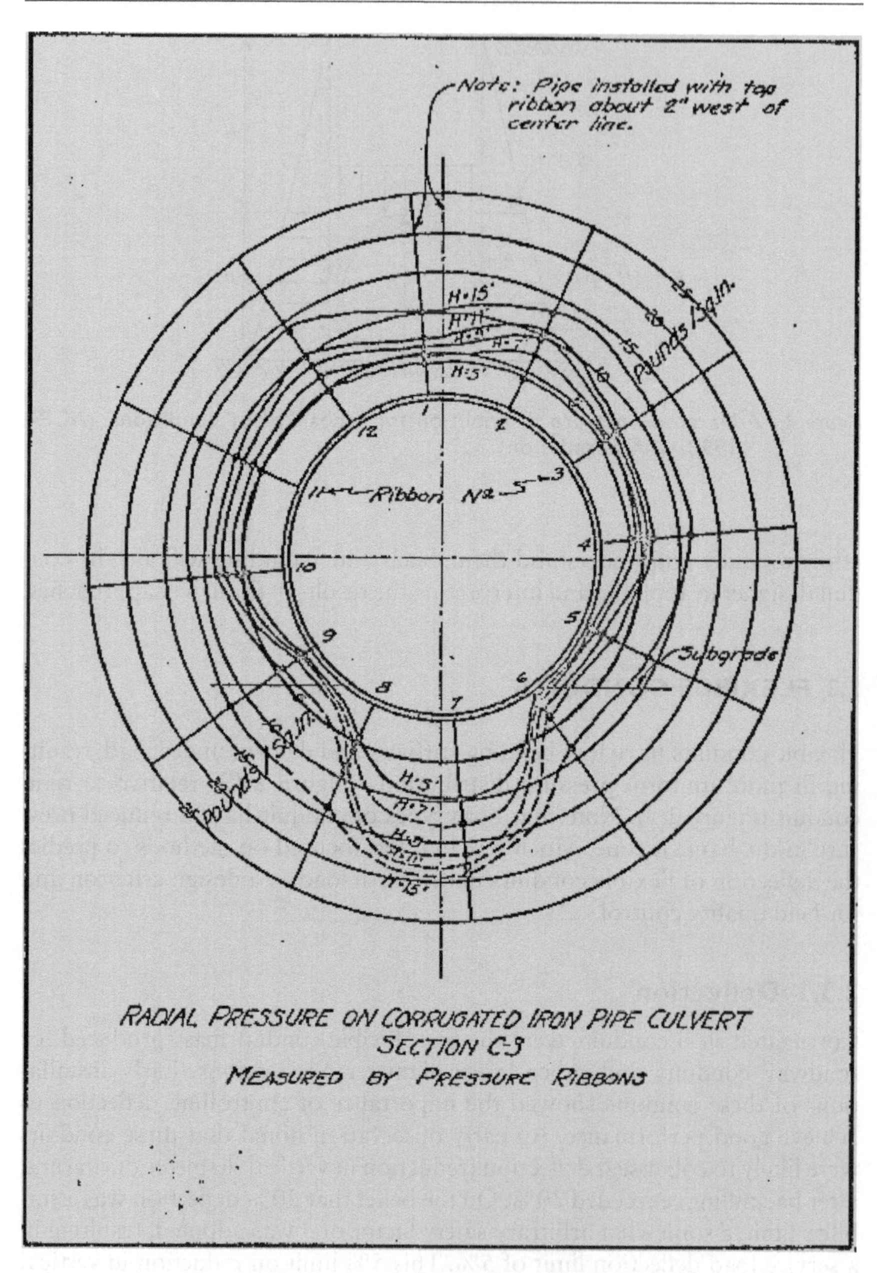

Figure 2.13 Radial pressures on flexible conduit. (Marston, 1930, with permission Iowa State Library Special Collections and University Archives)

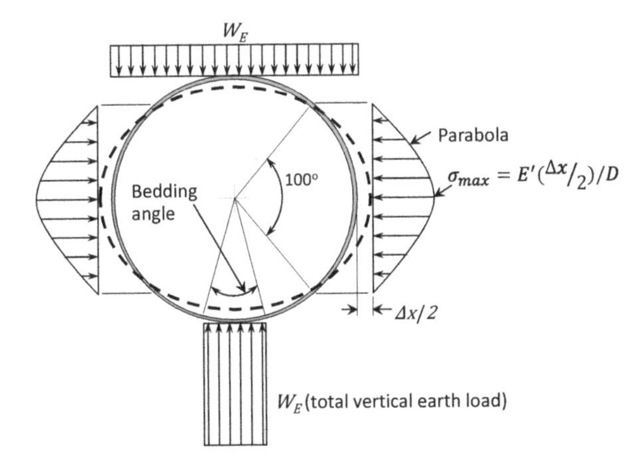

Figure 2.14 Spangler pressure distribution for flexible conduits.

that reduces conduit cost. Further, excessive deflection can result in leaking joints, which can affect long-term performance.

Spangler (1941) developed what is called the Iowa Formula or the Spangler equation for predicting deflection of flexible conduits subject to earth loads. The equation was developed by assuming a pressure distribution on the conduit (Figure 2.14) and then deriving an expression to predict the resulting change in horizontal conduit diameter. In simple terms, the Iowa formula can be written as a simple interaction equation:

$$\Delta x = \frac{\text{Load Term}}{\text{Conduit stiffness} + \text{Soil stiffness}}, \tag{2.4}$$

where Δx is the change in horizontal diameter. The full equation is presented and discussed in Chapters 5 and 7. An important result of Spangler's work was to demonstrate that soil stiffness is the dominant factor in limiting deflection of flexible conduits.

Features of the pressure distribution include

- W_E, the vertical earth load, assumed to be the soil prism load, was applied uniformly over the full diameter across the top of the conduit;
- the bedding pressure was applied under the conduit over the central portion of the conduit diameter defined by a bedding angle selected by the designer based on how the conduit will be installed; and
- a variable lateral pressure based on the amount of horizontal movement of the conduit as it deflects. The pressure is highest at the springline, where the conduit deflection is greatest. This distribution assumes all lateral pressure on the conduit results from horizontal deflection of the conduit into the soil. The lateral in situ pressure that would be present if the conduit did not deflect is not considered.

In current usage, the expression is typically used to predict the change in vertical diameter (see Section 7.7.1).

The soil parameter to describe the soil stiffness was called the modulus of passive resistance (e, psi/in., kPa/m), which was multiplied by the conduit radius to provide lateral pressure. The product $e\ R$ was later replaced with the modulus of soil reaction E' (psi, kPa), and the expression for maximum lateral pressure of

$$\sigma_{max} = E'(\Delta x / 2) / D, \tag{2.5}$$

where

σ_{max} = maximum horizontal stress at the side of the conduit, psi, kPa
E' = modulus of soil reaction, psi, kPa
Δx = total horizontal deflection of the conduit, in., m
D = diameter of conduit, m

Later work by McGrath (1998) suggested that the constrained modulus, M_s, a parameter derived from elasticity theory that varies with the confining stress is equivalent to E' and can be used in design. This is discussed in more detail in Section 4.3.

2.3.2 Hoop compression

To determine loads on flexible conduits, White and Layer (1960) proposed the hoop compression theory. This theory puts forth the concept that flexible conduits in embankment installations can be designed to carry the soil prism load as thrust at the springlines. This theory is inconsistent with elasticity theory for estimating the load carried by metal conduits but has been effectively applied to the design of flexible metal conduits, which have high hoop stiffness, with a factor of safety of 2. The corrugated metal conduit industry adopted the hoop compression theory as a primary design criterion. Controlling deflection remains important, but rather than designing backfill to control deflection, a deflection limit, often 5%, but sometimes higher, was established as a performance criterion. Most corrugated steel conduits will yield prior to reaching 5% deflection because steel is ductile and able to deform without failure. The hoop compression theory does not address the need to design flexible conduits for flexural capacity under some loading conditions, such as vehicular live load under low depths of fill. Some thermoplastic conduits, which are both flexible in bending and circumferentially compressible, required additional considerations discussed in Chapter 5.

2.3.3 Bedding support

Proper support to the bottom of the conduit is important to the performance of flexible conduits, just as for rigid conduits. Spangler's approach to bedding was to make the width of the bottom reaction a value selected by the designer. A narrow bedding will produce a more concentrated pressure on the invert of the conduit and thus increase the horizontal deflection. This was a somewhat simplistic approach and did not address the more significant effects of narrow bedding on the vertical deflection or the mitigation of that effect that might be achieved by proper placement and compaction of soil in the haunch zone. This is discussed in Chapter 4.

2.4 NONCIRCULAR CONDUITS

The key features of soil-conduit interaction described earlier apply to noncircular conduits in much the same way as for circular conduits. Arching will also influence loads on rectangular and arch shapes, and large-span metal conduits require lateral soil support to resist vertical loads and excess deformation. These structures are introduced in Chapter 3.

REFERENCES

ACPA (1998) *Concrete Pipe Handbook*, American Concrete Pipe Association, Irving, TX.

Heger, F.J. (1988) New Installation Designs for Buried Concrete Pipe, *Proceedings: Pipeline Infrastructure*, ASCE, Reston, VA, pp. 117–135.

Marston, A. (1930) The Theory of External Loads on Closed Conduits in Light of the Latest Experiments, *Bulletin 96*, Iowa State College of Agriculture and Mechanic Arts, Ames, IA.

Marston, A., and Anderson, A.O. (1913) The Theory of Loads on Pipes in Ditches, and Tests if Cement and Clay Drain Tile and Sewer Pipe, *Bulletin 31*, Iowa State College of Agriculture and Mechanic Arts, Ames, IA.

McGrath, T.J. (1998) Replacing E' with the Constrained Modulus in Buried Pipe Design, *Proceedings, Pipelines in the Constructed Environment*, J.P. Castronovo and J.A. Clark, Eds., American Society of Civil Engineers, Reston, VA.

NCPI (2017) *Vitrified Clay Pipe Engineering Manual*, National Clay Pipe Institute, Elkhorn, WI.

Spangler, M.G. (1941) The Structural Design of Flexible Pipe Culverts, *Iowa Engineering Experiment Station, Bulletin 153*, Iowa State College, Ames, IA.

Spangler, M.G. (1946) *Analysis of Loads and Supporting Strengths, and Principles of Design for Highway Culverts, Proceedings*, Vol. 26, Highway Research Board, Washington, DC.

Terzaghi, K. (1936) Stress Distribution in Dry and in Saturated Sand Above a Yielding Trap-Door, *Proceedings International Conference on Soil Mechanics*, Cambridge, MA, Vol. 1, pp. 307–311.

White, H.L., and Layer, J.P. (1960) The Corrugated Metal Conduit as a Compression Ring, *Proceedings, Highway Research Board*, Vol. 39, pp. 389–397.

Chapter 3

Conduit materials and general conduit behavior

Material performance issues such as corrosion, strength loss, and others are mentioned but not addressed in detail. National and local standards, and manufacturer's literature are available to address these issues.

Buried conduits interact with the surrounding soil. Designers make use of this interaction to create economical conduit installations. This chapter introduces the geometric and strength properties of conduits that contribute to that structural system and those that define the limiting behavior of the conduit. For this purpose, the presentation focuses on circular reinforced and nonreinforced concrete, metal, thermoplastic, and fiberglass conduits. This approach introduces the key properties related to soil-conduit interaction that apply to all other types of conduits. Conduits made of other materials and those manufactured in noncircular shapes and any variances due to shape or material will be introduced later.

The terminology used for conduit geometry was presented in Figure 2.1. Note that "crown" and "invert" identify the top and bottom inside the surface of the conduit. For noncircular sections, the maximum internal vertical dimension is the rise, and the maximum internal horizontal dimension is the span. Most structural behavior is calculated based on the mean diameter, $D_m = D_i + t$; however, vertical earth loads are calculated on the outside diameter, $D_o =$ or $D_i + 2\ t$.

3.1 CONDUIT MATERIALS

3.1.1 Concrete

Concrete conduits are manufactured in several closed shapes: circular, elliptical, and arch (Figure 3.1), as well as rectangular box sections and three-sided box sections without bottom slabs (Figure 3.2). Three-sided sections are also available with arched top slabs. Box sections can be precast or cast-in-place, while the other conduit shapes are all precast. Precast and cast-in-place arch sections (without bottom) are also used in some applications.

Concrete, an ancient building material used to manufacture conduits for centuries, is manufactured from a mixture of cement, sand, gravel, and water.

DOI: 10.1201/9780429162619-3

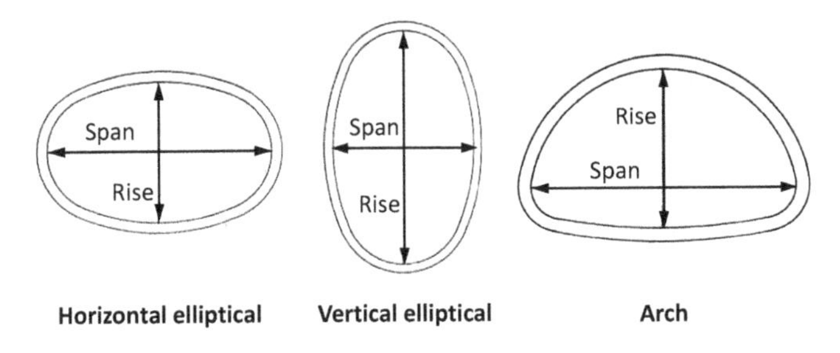

Figure 3.1 Noncircular concrete conduit shapes.

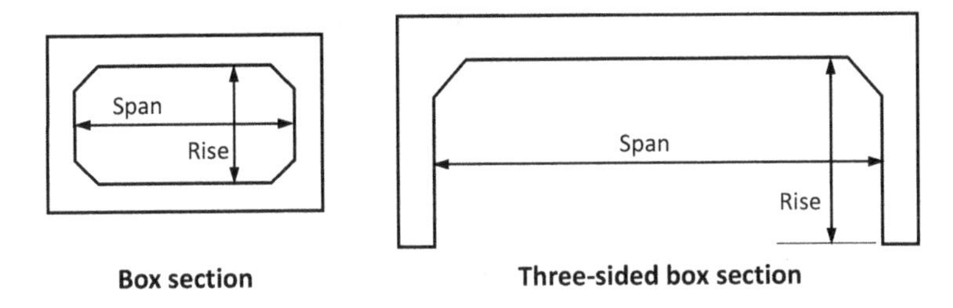

Figure 3.2 Rectangular concrete cross-sections.

Manufacturers also use additives to improve performance during mixing, placement, and curing, such as flow, rate of strength development, and durability. Concrete strength and stiffness properties vary with the concrete mix design. The key structural property is compressive strength. Most other design properties are estimated based on this one parameter. Concrete strength is variable with time. For design and quality control purposes, the strength is normally determined at the age of 28 days; however, precast concrete manufacturers generally desire rapid strength gain to allow early shipping and minimize required space for storage during curing. Thus, they often use mix designs, additives, and curing procedures to achieve design strength in a few days or less.

Properties – Key properties of concrete include:

- f'_c – Design compressive strength, ksi, (MPa)
- E_c – Young's modulus of elasticity, ksi, (MPa)

A number of equations have been proposed for estimating the modulus of elasticity of concrete. AASHTO LRFD suggests Equations (3.1) and (3.2) for concrete strengths up to 10 ksi:

$$E_c = k_1 \sqrt{f'_c} \tag{3.1}$$

or:

$$E_c = k_2\, w_c^{1.5} \sqrt{f'_c}, \tag{3.2}$$

where

w_c = unit weight of concrete, kcf, kg/m^3
k_1 = unit correction factor, 1,820 for US units, 4,730 for SI units
k_2 = unit correction factor, 33,000 for US units, 0.042 for SI units

Equation (3.1) is applicable to normal-weight concrete, while Equation (3.2) covers concrete with unit weights between 0.090 and 0.150 kcf (1,400 to 2,400 kg/m^3).

- f'_t – modulus of rupture (tensile strength). AASHTO (2020) takes the modulus of rupture as 0.24 $\sqrt{f'_c}$, ksi, (0.62 $\sqrt{f'_c}$, MPa), but the multiplier is taken between 0 and 0.38 (0 and 1.0 in SI units) depending on the application. In flexural design, the concrete tensile strength is typically conservatively assumed to be zero.
- f'_v – diagonal tensile (shear) strength. This was long taken as a multiple of the compressive strength:

$$f'_v = 0.063\sqrt{f'_c}\ \text{ksi}, \left(0.17\sqrt{f'_c}\ \text{MPa}\right) \tag{3.3}$$

However, modern design codes treat the diagonal tensile strength as a more complex variable dependent on the compressive strength, quantity of reinforcement, and strain in the reinforcement. These more complex design equations provide more accurate estimates of the diagonal tensile capacity of concrete members and will be discussed in Chapter 7.

Most concrete conduits are manufactured with cold-drawn wire reinforcement to resist tensile forces, which typically occur in the inside reinforcement at the crown and invert and in the outside reinforcement at the springlines. The reinforcement consists of the primary circumferential wires and longitudinal wires to hold the circumferential wires in place and to resist handling, shrinkage, and longitudinal in-ground forces. The combined longitudinal and circumferential reinforcement is often called a cage. The most common reinforcement configuration is two full circular cages, one near the inside surface and one near the outside surface (Figure 3.3).

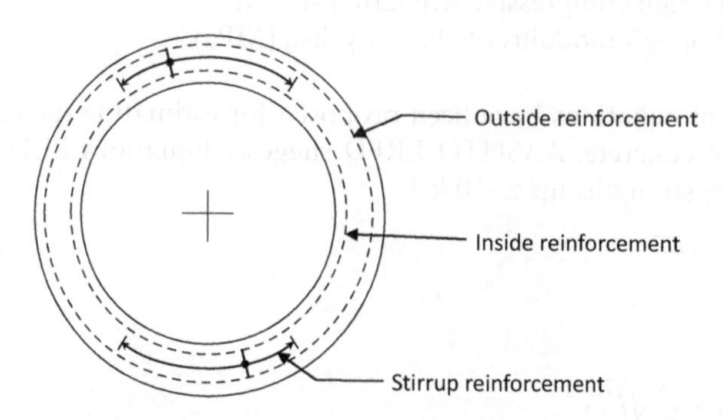

Figure 3.3 Typical reinforcement arrangement for concrete conduits.

Small-diameter conduits may have a single cage near the center of the wall or no reinforcement at all. While for a conduit in the ground, reinforcing only in the tension zones such as partial circular cages or elliptical reinforcement that is near the inside top and bottom surface and the outside surface at the springlines is possible and would reduce the total required reinforcement, this requires that the conduit be properly oriented when placed in the ground and it may not address possible tension locations during handling and installation. For conduits requiring substantial quantities of reinforcement, full circular cages are sometimes supplemented with partial circular cages in the tensile zones or elliptical cages.

Under some loading conditions, the diagonal tensile stresses create a limiting condition. In this case, the diagonal tensile capacity can be increased with stirrup reinforcement, which is oriented radially, extending from the inner to the outer reinforcement cage (Figure 3.3). Stirrups are typically installed only in the invert and crown region of the conduit which are the areas of excess diagonal tensile stress; however, this requires installation of the conduit with the proper orientation. Stirrups must be anchored near the inner and outer reinforcement. Stirrups can be individual reinforcement bars wrapped around the inner and outer circumferential reinforcement but are most often supplied as mats that can be inserted into the primary reinforcement.

In design, concrete is typically treated as a linear elastic material with a constant modulus of elasticity; however, the actual behavior (e.g., at high stress or after cracking) can be more complex. Figure 3.4 presents a typical stress-strain curve for concrete (solid line) and two simplified curves (t_1, t_2) representing different times after loading as used in some computer programs.

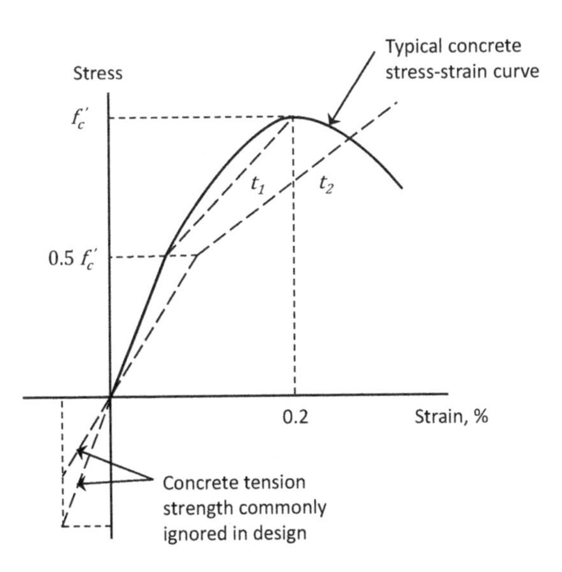

Figure 3.4 Concrete stress-strain curve and simplified model. (Heger and Liepins 1985, Authorized reprint from 1985 Proceedings ACI Journal, Vol. 82, No. 3)

Features of Figure 3.4 and their use in design include the following:

- While the strength is variable with age, the strain at peak stress is reasonably consistent at 0.2%. This value is assumed constant for standard design calculations.
- Stress-strain behavior is reasonably linear at low stresses and less so as the stress approaches the peak. The simplified modeling assumption shown in Figure 3.4 is to assume a constant modulus of elasticity up to 50% of peak stress and a reduced constant modulus to the peak stress at a strain of 0.2%.
- Concrete will creep under sustained loading. If this is a design consideration, a lower modulus can be used (t_2). For typical applications, creep in concrete is ignored.
- Concrete is typically assumed to have no tensile strength.

Design – Concrete conduits are designed for three ultimate strength conditions:

- Flexural strength – flexural strength is controlled by the reinforcement yield strength or by the concrete compressive strength. Failure by yielding is preceded by flexural cracking (Figure 3.5), which is acceptable if not excessive. Failure by compression occurs rarely but can occur at the springlines from compressive thrust due to vertical earth load and high bending moments.
- Shear strength – shear forces create diagonal tensile stresses in concrete members. Excessive shear force results in diagonal cracking through the conduit wall (Figure 3.6).
- Radial tensile strength – radial tension occurs when the inside reinforcement is under tension, which creates a tensile stress in the concrete, restraining the reinforcement from straightening (Figure 3.7). Radial tensile failure is often called "slabbing" as the cover concrete pulls away from the main body of the conduit wall.

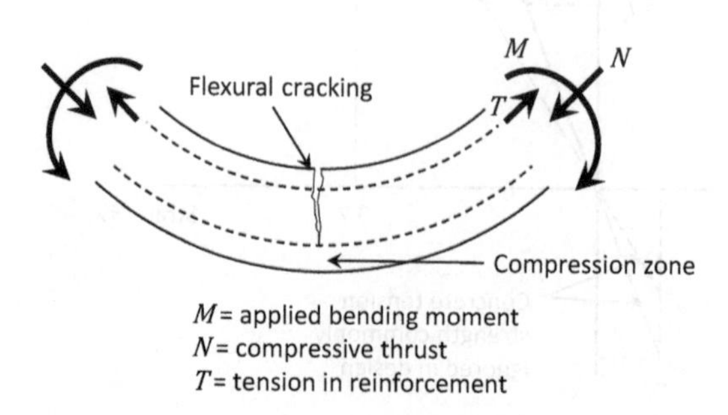

Figure 3.5 **Reinforcement to resist bending moments in concrete conduits.**

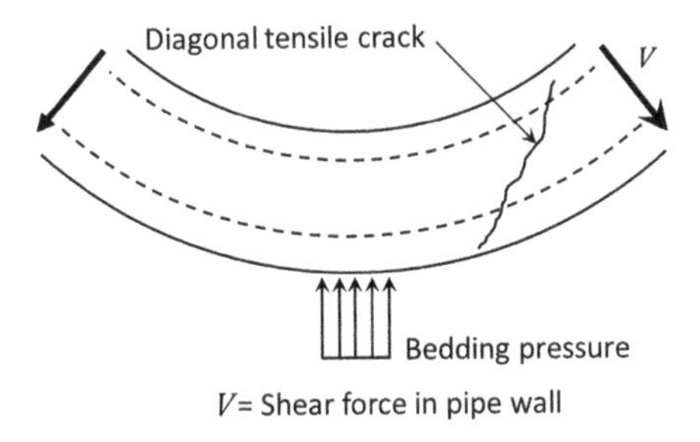

Figure 3.6 Diagonal tension in concrete conduits.

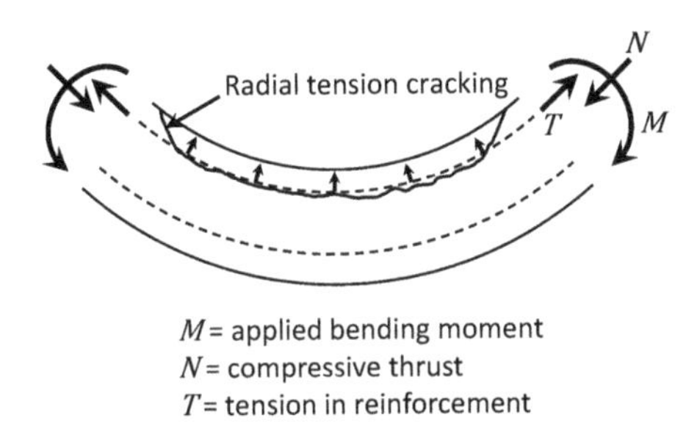

Figure 3.7 Radial tension in concrete conduits.

A service load design condition for concrete conduits is the allowable width of flexural cracks. This limit is commonly set at 0.01 in. (0.25 mm or 0.3 mm, depending on the code).

Concrete is a durable material except in the presence of sulfates or low-pH fluids or soils. In some conditions freezing and thawing cycles can cause deterioration of the concrete.

3.1.2 Metal

Metal conduits are manufactured from steel or aluminum and are provided in many shapes, some of which are shown in Figure 3.8, and sizes. Current US specifications in the AASHTO LRFD Bridge Design Specifications (AASHTO, 2020, called AASHTO LRFD) provide separate design procedures for several of the categories of these structures, including the following:

- Metal Pipe, Pipe Arch, Arch Structures, and Steel Reinforced Thermoplastic Culverts (Article 12.7)
- Long-Span Structural Plate Structures (Article 12.8)
- Deep Corrugated Structural Plate Structures (Article 12.8.9)
- Structural Plate Box Structures (Article 12.9)

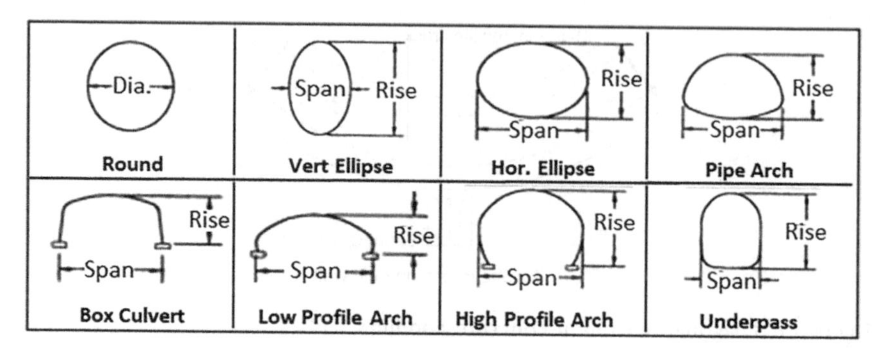

Figure 3.8 Metal conduit shapes. (NCSPA, 2018, adapted with permission)

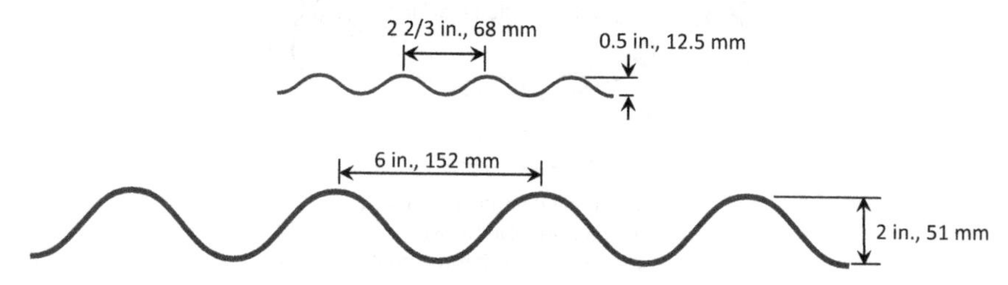

Figure 3.9 Sample corrugated profile configurations for metal. (NCSPA, 2018, adapted with permission)

Several of the aforementioned metal conduit design methods were developed at the same time as our understanding of soil-conduit interaction was evolving and before computer analysis and design were available. As a result, these methods are semiempirical and prescribe dimensional limits and structural features to keep behavior within the limits of understanding at that time. More recently, computer modeling is used for design of the larger structures and design is based on expected behavior, allowing some of the limits of the empirical methods to be removed. This is discussed in more detail in Chapter 7.

Metal gravity flow conduits are manufactured of steel or aluminum with corrugated (Figure 3.9) or spiral rib (Figure 3.10) profiles. Profile designs both deeper and shallower are available.

Steel is assumed to provide a linear response up to the yield strength (F_y), with a modulus of elasticity (E_s) of 29,000,000 psi (200 MPa). Aluminum does not have a clearly defined yield point but is assumed linear in the service load range with a modulus of elasticity of 10,000,000 psi (70 MPa). Typical idealized stress-strain curves for steel and aluminum are presented in Figure 3.11. Cold-worked steel, such as wire reinforcement for concrete conduits, is steel that has been stressed into the strain-hardening region and has a stress-strain curve similar to aluminum with a higher yield stress than mild steel.

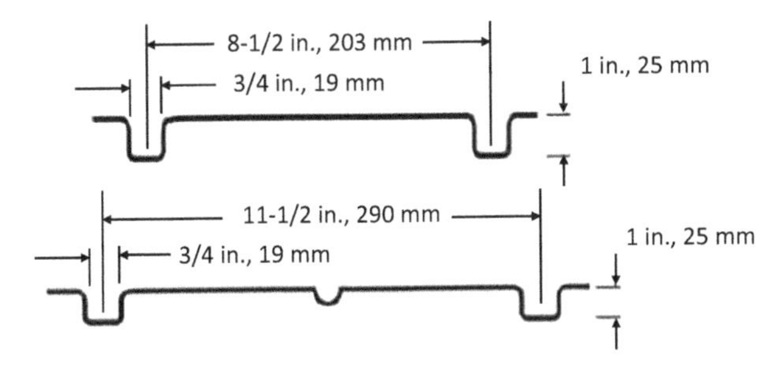

Figure 3.10 Sample spiral rib profile configurations for metal conduits. (NCSPA, 2018, adapted with permission)

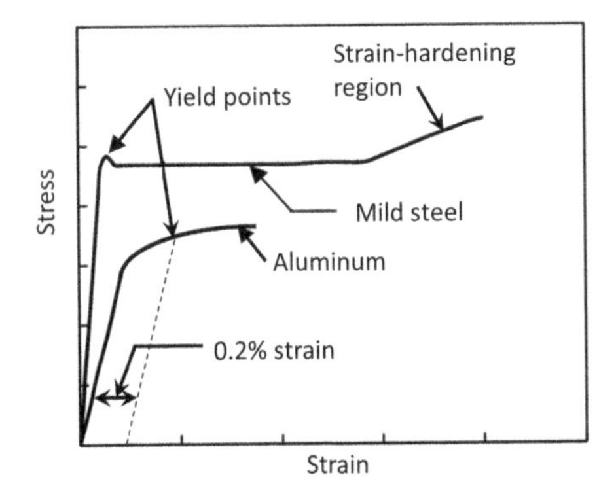

Figure 3.11 Idealized stress-strain curves for steel and aluminum.

Figure 3.11 shows the following behavior:

- Steel exhibits linear behavior to an initial peak called the yield strength (F_y) followed by a near-zero modulus "plastic region" where little or no increase in stress results from an increase in strain, and then a strain-hardening region where increasing stress is required to further increase the strain up to an ultimate strength (F_u) which is the maximum stress the section can carry.
- Aluminum does not have a clearly defined yield point such as steel. For design, the yield stress is often taken at the point where the stress-strain curve deviates from linear by 0.2% (the 0.2% offset stress). The yield point of aluminum may be greater or less than that of steel.

Steel and aluminum conduits can both be stressed beyond their yield points without causing a failure, provided the soil support prevents excessive deformation that could lead to collapse. It is not uncommon for yielding to occur in buried conduits that provide good service. Figure 3.12 shows the stress profile in metal conduits at the onset of yielding in flexure and at the

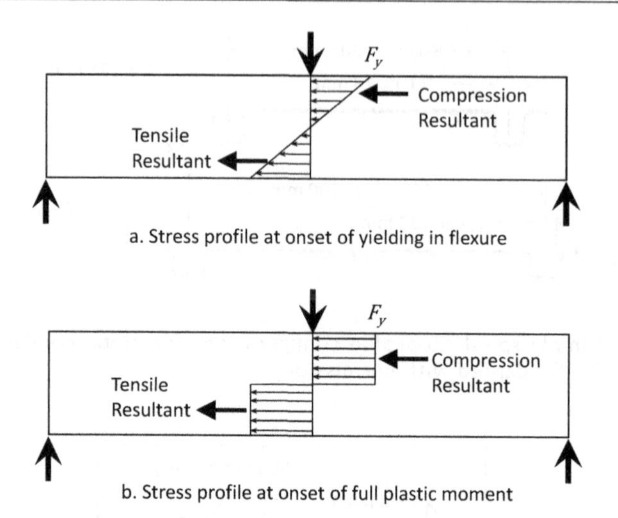

Figure 3.12 Stress profiles for steel at yield stress and plastic moment capacity.

full plastic moment. Design for steel conduits in flexure is sometimes based on using the plastic moment capacity as a strength limit state.

AASHTO (2020) sets a flexibility limit (*FF*, Eq. (3.4)) to provide a minimum stiffness that will limit conduit deformation during backfilling. The flexibility factor is presented in AASHTO LRFD as a strength limit state but is better considered as a service limit state.

$$FF = \frac{D_i^2}{EI}, \tag{3.4}$$

where

FF = flexibility factor, in./kip, m/kN
D_i = diameter or span of conduit, in., m

The terms in Equation (3.4) are often presented in different units. The calculation must be made in consistent units. AASHTO sets maximum limits on the flexibility factor based on the conduit shape, material, and corrugation type and thickness. The flexibility factor is discussed further in Section 3.2.1

AASHTO requires no design check for a service limit state on metal conduits; however, deflection, i.e., change in diameter after backfilling, is often monitored and must be limited.

Strength limit states vary with the type of structure but include the following:

- Wall area – limits the average compression stress due to axial thrust in the conduit wall

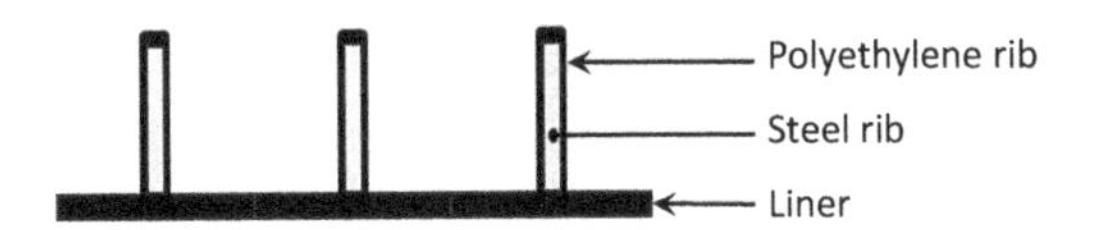

Figure 3.13 Steel-reinforced thermoplastic conduit wall profile.

- Buckling strength – maintains stability of the conduit wall under compression loading
- Seam failure – limits the compression across folded and bolted seams
- Flexural capacity – metal box sections and structures with deep corrugations, 6 in. (150 mm) and deeper, are evaluated for flexural resistance. The AASHTO LRFD design procedures for long-span conduits (Article 12.8) do not explicitly check flexural capacity. If the prescribed limits in the design section are followed, flexural capacity is assumed to be adequate for these structures.

Steel is susceptible to corrosion (rust) in wet or low-pH conditions and when embedded in soils with low resistivity. Corrosion resistance for corrugated steel conduits is mitigated with the use of galvanized (zinc-rich) or polymeric coatings or aluminized steel. Users must investigate the benefits and limitations of these products and other products to enhance corrosion resistance based on site conditions and desired service life before specifying.

Aluminum has generally good corrosion resistance in typical in-ground conditions, but as with steel, specific site conditions must be evaluated prior to specifying.

Steel-reinforced thermoplastic conduit – Steel-reinforced thermoplastic conduit is a composite profile designed under AASHTO LRFD Article 12.7 for metal conduits but is quite different from a metal conduit. The conduit wall construction is shown schematically in Figure 3.13. The main reinforcement element is the steel rib, which is held in place and protected from corrosion by polyethylene encasement. The polyethylene liner provides the conduit shell. The use of pressure-rated polyethylene resins eliminates the need for stress crack tests, as stress crack resistance is provided through the pressure rating process (see Section 3.1.3.1).

3.1.3 Plastics

Plastic conduits include thermoplastic and thermoset conduits manufactured with a wide range of materials with a similarly wide range of stiffness and strength properties. The most common thermoplastic materials for gravity flow conduits are polyvinyl chloride (PVC) and high-density polyethylene (HDPE), but the use of corrugated polypropylene conduit is increasing. Thermoset conduits are generally manufactured from fiberglass, also called GRP (glass-reinforced polymer) or GFRP (glass fiber–reinforced

Figure 3.14 Wall profiles for thermoplastic conduits. (McGrath et al. 2009, Figure B-4, p. B-5, Copyright National Academy of Sciences. Reproduced with permission of the Transportation Research Board)

polymer). Thermoplastic conduits manufactured for gravity flow applications typically have a profile wall configuration (Figure 3.14), while thermoset conduits are typically manufactured with a solid rectangular wall section. Large-diameter fiberglass conduits have been manufactured with exterior ribs to improve stiffness and flexural capacity.

3.1.3.1 Thermoplastics

Thermoplastics can be heated and cooled within defined ranges without changing the engineering properties. This allows thermoplastic resins used to manufacture conduits to be produced, usually in pellet form, and shipped in convenient bulk containers. Manufacturers heat and form the plastic into an end product, usually by extrusion, which is then cooled for delivery to users. There are many properties of interest concerning both processing parameters and mechanical properties when selecting a thermoplastic. Cell classification systems are used to assist users in identifying suitable resins for projects. Two of these are ASTM D1784 *Standard Classification System and Basis for Specification for Rigid Poly(Vinyl Chloride) (PVC) Compounds and Chlorinated Poly(Vinyl Chloride) (CPVC) Compounds* and ASTM D3350 *Standard Specification for Polyethylene Plastics Pipe and Fittings*

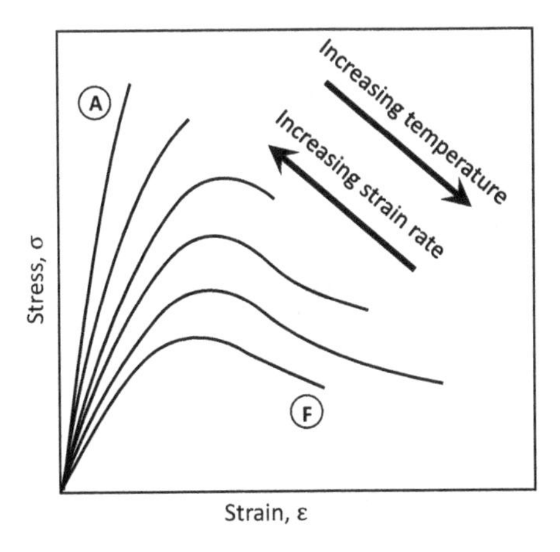

Figure 3.15 Generic stress-strain curves for thermoplastics.

Materials. These standards do not address the long-term performance of resins, and thus additional testing is required, as will be discussed next.

General behavior of thermoplastics – Thermoplastics respond to loads differently with temperature and time under load. The varying slopes of the stress-strain curves in Figure 3.15 imply a change in modulus of elasticity; however, the change is due to creep or relaxation (explained later in this section). Thus, the slope of the stress-strain curves should be considered an apparent modulus for the specific load rate and temperature at the time of testing.

Figure 3.15 shows that as the test temperature increases, the stress-strain curves move from Curve A to Curve F – that is, the apparent modulus of elasticity and strength (peak stress) decreases. Conversely, as the strain rate (rate of loading) increases (shorter time periods), the stress-strain curves move from Curve F to Curve A, and the apparent modulus of elasticity and strength increases.

This stress-strain behavior can be demonstrated with a model of springs and dashpots, much like the springs and shock absorbers in an automotive suspension system. A spring represents a simple elastic system in which an applied force produces an immediate fixed displacement, and upon removal of the force, the spring returns to its original length. A dashpot will deform only as a function of time under load, and when the load is removed, the dashpot will remain in the deformed position. Figure 3.16 represents such a system that describes the behavior of plastics. Although more complex and precise systems are available, this is adequate to describe the behaviors we are interested in for buried, gravity flow conduit installations.

The following occurs when this model is subject to a constant tension or compression load:

- Element A, a spring, deforms immediately and remains deformed the same amount under the constant load. This is called the elastic response, and when the model is unloaded, it rebounds immediately to its original length.

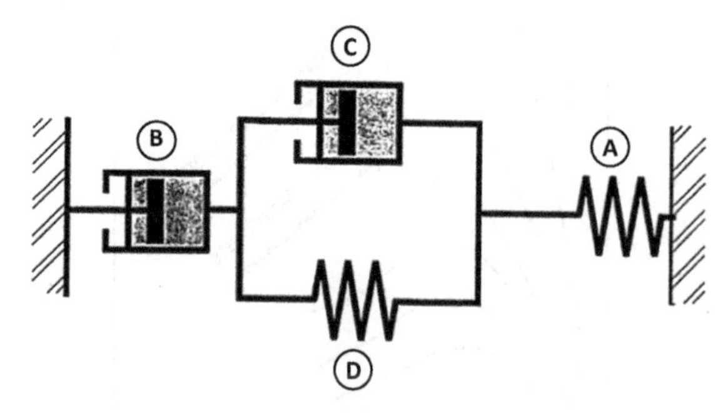

Figure 3.16 Four-element, time-dependent model for plastic stress-strain behavior.

- Element B does not deform immediately but deforms slowly with time and stays in the deformed position when the load is removed. This represents permanent plastic deformation.
- Elements C and D work in tandem. When the model is loaded, there is no immediate deformation. Over time, Element C deforms at a decreasing rate as the resulting deformation gradually deforms Element D, reducing the load on Element C. When C and D are unloaded, the compression in the spring will gradually cause Element C to move back toward its undeformed position at an ever-slowing rate.
- The deformation in C and D plus the deformation in B represents creep: ongoing deformation under constant load.
- The rebound in Elements C and D after unloading represents strain recovery.

The response to a constant load is creep – an increasing deformation over time, but at a decreasing rate. A plastic water conduit under constant internal pressure responds with creep in tension. Soil pressure around the perimeter of a conduit creates a creep stress in compression. Conduits under internal pressure are outside the scope of this book.

The converse of creep is relaxation, where a specimen is held at a constant deformation. When the model in Figure 3.16 is held at a fixed deformation, the load initially deforms Element A, but over time, Element B deforms permanently, and Element C deforms at a decreasing rate as Element D takes up load. This results in an overall reduction in load as Element A returns to its original length. Relaxation is the load condition of a conduit held at a constant deflection between two parallel plates and is similar to the condition of an installed conduit held in shape by soil embedment. As in the creep condition, when unloaded, there is an initial rebound (Element A), a time-dependent rebound (Elements C and D), and a permanent deformation (Element B). Creep and relaxation are discussed in more detail here and later when soil-conduit interaction is introduced.

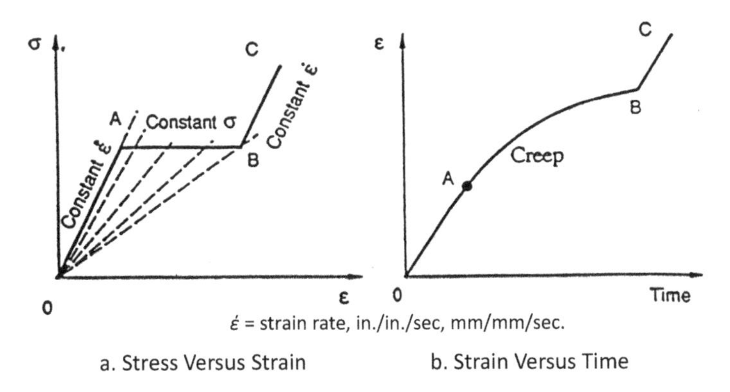

Figure 3.17 Creep in thermoplastics. (McGrath et al., 1994, Reprinted with permission, Copyright ASTM International, www.astm.org)

Creep – Creep can be demonstrated by loading a series of specimens at different rates. Figure 3.17 shows stress versus strain for varying strain rates, $\acute{\varepsilon}$.

In Figure 3.17a, the dotted lines represent separate specimens loaded at different strain rates (in./in./sec, mm/mm/sec). The higher the strain rate, the steeper the curves, resulting in a higher apparent modulus of elasticity. If the test at the highest strain rate is held at the load associated with Point A, it will continue to deform in creep, i.e., the deformation will increase under constant stress, and the specimen deformation moves from Point A to Point B. If the test is restarted at Point B at the original strain rate, the slope of the stress-strain curve (B to C) will be the same as the initial portion of the test (0 to A), not at the slope of the curve it intersects at Point B. The results of the same test A are plotted as strain versus time in Figure 3.17b. In this figure, the increase in strain under constant stress from Point A to Point B is clearly visible. This demonstrates a principle of linear viscoelasticity (at stress levels typically encountered in properly buried conduits at moderate stress levels) that the apparent modulus of elasticity is the same for a given strain rate and temperature regardless of the prior history of the specimen.

Performance at stress levels varying from low to high is demonstrated schematically in Figure 3.18 by showing strain versus time for a series of specimens placed under creep loading with a range of stresses.

Figure 3.18 shows the following behavior:

- As the stress level increases, the rate of creep strain increases with time. Compare Curve $\sigma 1$ with Curve $\sigma 7$.
- Under low stress, Curve $\sigma 1$, the strain increases with time and does not accelerate, and the specimen will not fail for an indefinite period.
- As the applied stress level increases, the strain rate increases with time and leads to failure in successively shorter periods of time.

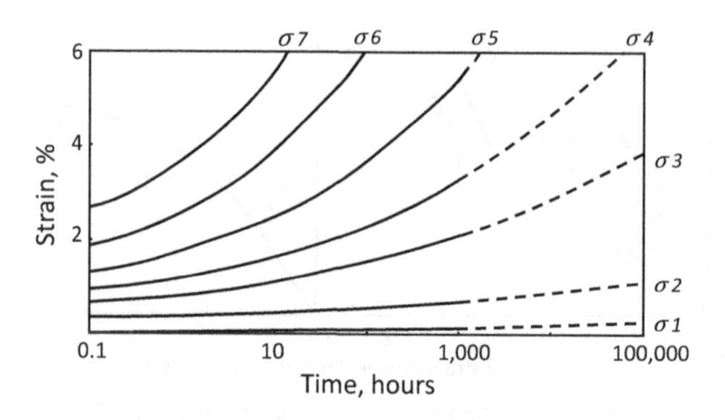

Figure 3.18 Strain versus time under constant stress.

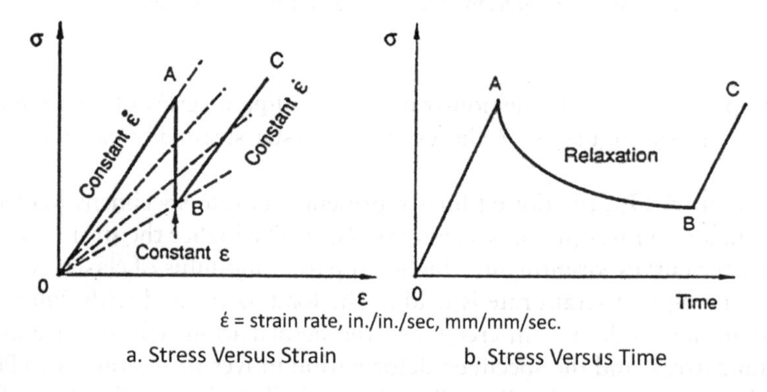

Figure 3.19 Relaxation of plastics. (McGrath et al., 1994, Reprinted with permission, Copyright ASTM International, www.astm.org)

The key point is that short-term tests do not provide sufficient information to design for long-term applications. This is discussed further next.

Relaxation – Relaxation can also be demonstrated by looking at stress versus strain and stress versus time curves (Figure 3.19).

Figure 3.19a demonstrates the same linear viscoelastic principle that was shown in Figure 3.17 for creep. After specimen loading is stopped at Point A, the deformation is held constant, and the stress drops. When loading is restarted at the same rate, the slope of the curve from Point B to C matches the original slope from A to B; thus, provided the polymer was not stressed beyond the linear viscoelastic range, the prior history does not affect the response to an instantaneous load condition. Figure 3.19b shows the same test plotted as stress versus time, demonstrating decreasing stress at an ever-slowing rate.

Figure 3.20 demonstrates the relaxation behavior of two corrugated HDPE conduits loaded at different rates and held at 10% deflection for almost two years while being subjected to periodic incremental load tests.

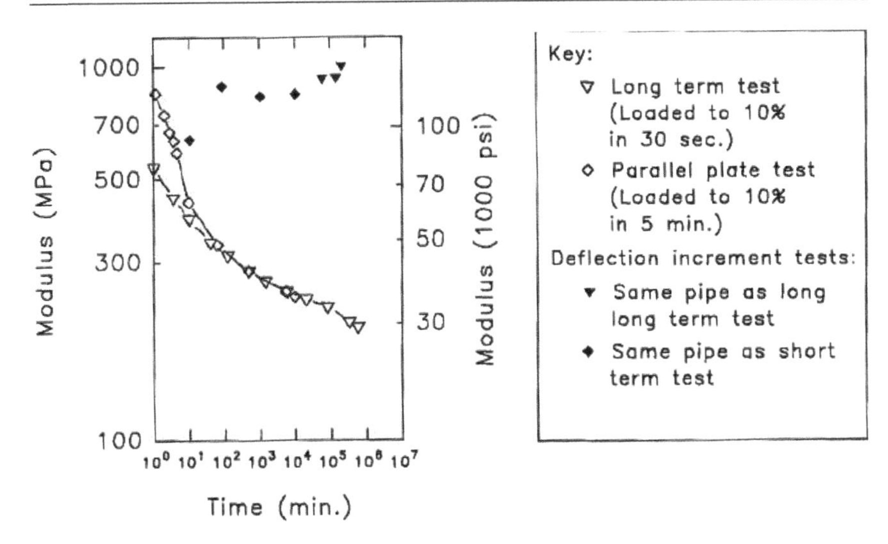

Figure 3.20 Modulus versus time for PE conduits at 10% deflection. (McGrath et al., 1994, Reprinted with permission, Copyright ASTM International, www.astm.org)

The slowly loaded conduit shows a lower initial apparent modulus, but the long-term apparent modulus of both conduits is the same. The conduits always respond to a short-term load with a short-term apparent modulus; thus, for a shallow buried HDPE conduit, the apparent modulus under a live load condition will be the short-term modulus.

Strength – Design with plastics is most often concerned with establishing tensile strength. Procedures have been developed to determine the maximum tensile stress that may be applied for the appropriate design life of the plastic product. For pressure conduits, this is most often achieved by loading specimens to varying stress levels and waiting for failure. These tests are typically conducted to produce failures at times up to at least 10,000 hours. The resulting data is then extrapolated to the desired design life of the product. Details of the test are found in ASTM D1598 *Standard Test Method for Time-to-Failure of Plastic Pipe Under Constant Internal Pressure*, and a method of analysis in D2837 *Standard Test Method for Obtaining Hydrostatic Design Basis for Thermoplastic Pipe Materials or Pressure Design Basis for Thermoplastic Pipe Products*, which linearizes and extrapolates the data on a log-log plot to the desired design life. A sample strength regression curve for a PVC resin is shown in Figure 3.21. A comparable standard for fiberglass conduit is ASTM D2992 *Standard Practice for Obtaining Hydrostatic or Pressure Design Basis for "Fiberglass" (Glass-Fiber-Reinforced Thermosetting-Resin) Pipe and Fittings.*

Short-term burst tests of PVC conduits subjected to long-term hydrostatic tests indicate that the short-term burst strength does not change with time under stress. Figure 3.22 shows tests where pipe specimens were held under

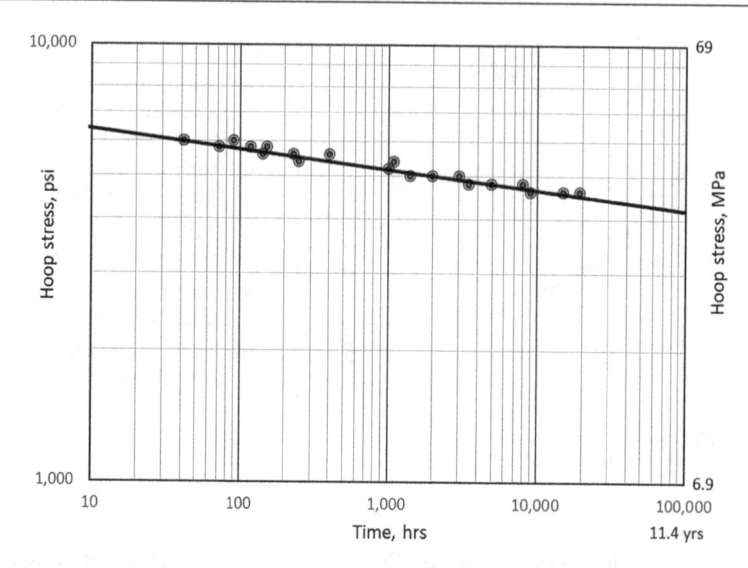

Figure 3.21 Strength regression of PVC. (Uni-Bell, 2012, reproduced with permission)

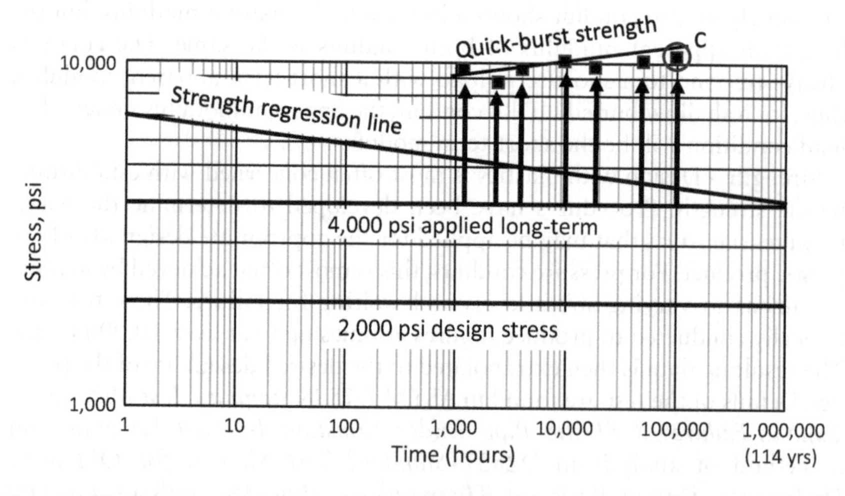

Figure 3.22 Short-term burst tests on long-term hydrostatic test specimens. (Uni-Bell, 2012, reproduced with permission)

constant internal pressure to a stress just below the long-term strength and then subjected to quick burst tests, which showed strengths similar to the short-term strength. The test marked "C" was just short of its projected long-term strength when tested. This is similar to how thermoplastics under relaxation respond to short-term loads with the short-term modulus (Figure 3.20).

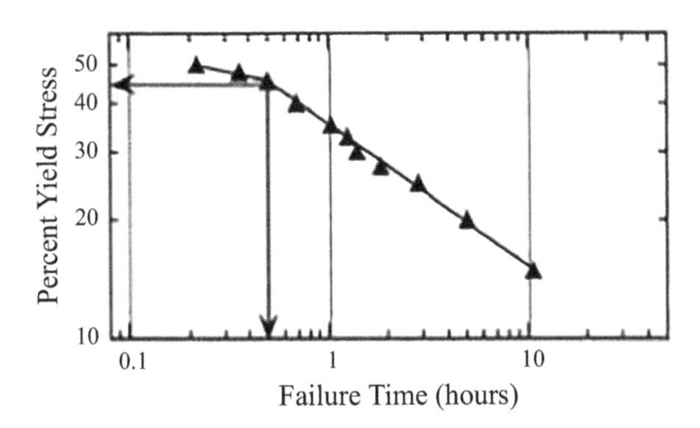

Figure 3.23 NCLS test results for HDPE resin. (Hsuan and McGrath, 1999, Figure 12, p. 16, Copyright National Academy of Sciences, reproduced with permission of the Transportation Research Board)

Hydrostatic pressure testing addresses conduit strength under creep tensile loading, while gravity flow conduits are subjected to compression from external earth load and bending from deflection. Compression stresses are evaluated using the tensile strength but also must be evaluated for resistance to buckling. Strain levels are often used to define limit states.

Under typical burial conditions for gravity flow conduits, where relaxation is the controlling behavior, it is problematic to determine a strength limit, as it is difficult to produce failures in short-term tests. PVC resins approved for use in buried conduit applications have high strain limits, and failures due to deflection should not occur at typically accepted deflection levels. However, some plastics, such as polyethylene, will transition to a brittle failure mode with a more rapid loss of strength than predicted by a straight-line regression. The change in slope of the strength regression curve is often referred to as a "knee." ASTM F2136 *Standard Test Method for Notched, Constant Ligament-Stress (NCLS) Test to Determine Slow-Crack-Growth Resistance of HDPE Resins or HDPE Corrugated Pipe* was developed to evaluate this issue for gravity flow conduits. The test consists of notching a tensile specimen, placing it under constant stress, and immersing it in an accelerating environment. Figure 3.23 presents results from one series of tests and shows a change in slope (knee) at about 1 hr. Results from this test were calibrated against field data to set a failure criterion that assures good long-term performance. The knee in hydrostatic regression plots is more pronounced than that in the NCLS data.

3.1.3.2 Thermoset plastics

Thermoset plastics are formed into a final shape at the time of manufacture. Glass fiber reinforcement is saturated with a liquid resin and cured to form a conduit. Fibers are placed by winding continuous filaments onto an interior mandrel or chopped fibers introduced into a spinning external mold

Table 3.1 Mechanical properties of fiberglass resins and glass fibers (condensed from AWWA, 2014).

Property	*Units*	*Resin*	*Glass Fibers*	*Conduit*	*Units*	*Resin*	*Glass Fibers*	*Conduit*
Tensile strength	10^3 psi	9.0–13	250–350	2.0–80.0	MPa	62–90	1,725–2,400	14–550
Tensile Modulus	10^6 psi	0.4–0.6	10–11	0.5–5.0	GPa	2.8–4.1	69–76	3.5–34.5

Source: reprinted with permission. M45 Figberglass Pipe Design, 3rd edition, American Waterworks Assn.

(centrifugal casting). The *Fiberglass Pipe Design Manual, American Water Works Association Manual of Supply Practices M45, 3rd Edition* (AWWA, 2014) provides details on materials, manufacture, and design of fiberglass conduits. Some typical material properties of fiberglass conduit and its components are presented in Table 3.1

Several thermoset resins are used in the manufacture of fiberglass conduit, mostly to achieve varying levels of corrosion resistance. These resins generally fall under the category of polyester or epoxy. Gravity flow conduits are primarily manufactured with polyester resins. An interior liner using a different resin, often with a lower modulus of elasticity, can be introduced to improve corrosion resistance to fluids carried by the conduit.

Table 3.1 shows that the strength and stiffness of fiberglass conduits are derived almost completely from the glass fibers. The resin serves to bind the composite together and provide corrosion resistance, except in chopped fiber wall constructions. In this latter case, forces must transfer from fiber to fiber by passing through the lower modulus resin, and hence, lower properties result.

Fiberglass conduits show creep and relaxation behavior under constant load/displacement, as well as strength regression under constant stress; however, since the bulk of the strength and stiffness is derived from the glass fibers, which do not exhibit significant viscoelastic behavior, the change over time of the modulus and strength is much less than for thermoplastics.

Fiberglass conduits are used for many applications, including municipal works, oil fields, and chemical plants. Their corrosion resistance to most environments is high but varies as a function of the resins used in manufacture. Fiberglass conduits intended for sewer applications are tested in a deflected shape while exposed to an acid environment. See ASTM D3681 *Standard Test Method for Chemical Resistance of "Fiberglass" (Glass–Fiber–Reinforced Thermosetting-Resin) Pipe in a Deflected Condition.*

3.2 RING THEORY – PROPERTIES OF ROUND CONDUITS

The materials introduced briefly in Section 3.1 must be manufactured into conduits. It is useful to look at the equations that describe the behavior of conduits when subjected to various simple load conditions, as these simple load conditions are often used for quality control tests of conduit and are sometimes modified to provide simplified design equations for field applications. For the purposes of this section, we discuss only round conduits.

3.2.1 Conduit characterization

Soil-conduit interaction first requires an independent understanding of the soil and conduit properties that contribute to in-ground behavior. The key conduit parameters are the hoop (membrane compressibility) stiffness and bending (flexural) stiffness.

Hoop stiffness, PS_H, is the circumferential extension/compression that occurs when a round conduit is subjected to a unit internal or external pressure, as defined in Equation (3.5) and Figure 3.24.

$$PS_H = \frac{p}{\Delta D / D_m} = \frac{EA}{R_m}, \tag{3.5}$$

where

PS_H = hoop stiffness – change in conduit diameter due to a unit applied internal or external pressure on the conduit, lbs/lineal in./in., kN/lineal m/m
R_m = radius to centroid of conduit wall, in., m
D_m = diameter to centroid of conduit wall, in., m
p = applied internal or external pressure, psi, kPa
ΔD = change in conduit diameter, in., m

Bending stiffness, PS_B, reflects the change in conduit diameter resulting from two diametrically opposed loads, as defined in Figure 3.25.

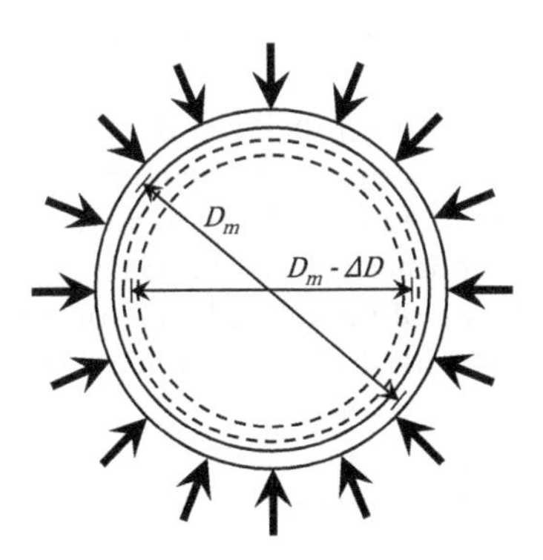

Figure 3.24 Conduit hoop stiffness.

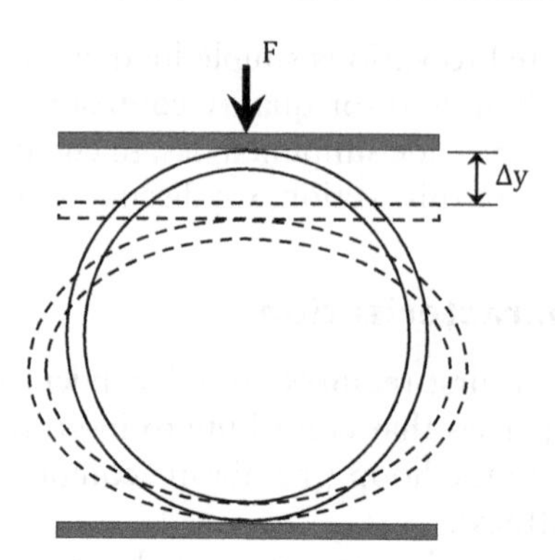

Figure 3.25 Parallel-plate test for conduit bending stiffness.

In US practice, this is computed as the ratio of the applied force to the resulting deflection:

$$PS_{B-US} = \frac{F}{\Delta_y} = \frac{EI}{0.1488\, R_m^{\,3}}, \tag{3.6}$$

while many international standards define stiffness using only the pertinent geometric and conduit material parameters:

$$PS_{B-I} = \frac{EI}{D_m^{\,3}}, \tag{3.7}$$

where

$PS_{B\text{-}US}$ = conduit bending stiffness as defined in ASTM standards, lbs/lineal in./in., kN/lineal m/m
$PS_{B\text{-}I}$ = conduit bending stiffness as used in many world standards, lbs/lineal in./in., kN/lineal m/m
Δy = change in vertical diameter resulting from force F, in., m
F = applied load causing change in diameter, lb/in., kN/m

The definitions of $PS_{B\text{-}US}$ and $PS_{B\text{-}I}$ spell out the units as lb/lineal in./in., (kN/lineal m/m) to emphasize that the stiffnesses relate load to deformation and are not a stress as might be implied if expressed as psi (kPa). Also, $PS_{B\text{-}US}$ and $PS_{B\text{-}I}$ differ only by a numerical constant: $PS_{B\text{-}US} = 53.76\ PS_{B\text{-}I}$, and thus are identical for the purpose of modeling behavior. Some analytical methods for conduit-soil interaction use EI/R^3 to define stiffness. Again, this is just a different version of the same parameter.

Another form of defining stiffness is the flexibility factor, FF (Eq. (3.4)), which is a function of EI/D^2. Equation (3.6) can be rearranged in terms of EI/D^2 as

$$\frac{F}{\Delta y / D_m} = \frac{53.76}{FF} = 53.76 \frac{EI}{D_m^2}. \tag{3.8}$$

Thus, the flexibility factor EI/D^2 is a function of the force to cause a unit deflection expressed as a fraction of the diameter, while the conduit stiffness equation is expressed as the force required to cause a unit displacement (in., mm). The consequence of this is that if the conduit stiffness is held constant as the diameter increases, the flexibility factor will continue to decrease (Figure 3.26). This figure does not show SI units; however, the intent of the graph is clear that conduits with diameters greater than about 27 in. could be less stiff if using the flexibility criterion.

The significance of the bending stiffness and hoop stiffness parameters, and the rationale for some of the distinctions that are made between various types of conduits during design can be demonstrated with calculations for some typical conduit dimensions and properties. Table 3.2 presents the properties used, with EI/R^3 representing bending stiffness to be consistent with the work presented in Chapter 4.

When the stiffnesses are plotted as hoop stiffness versus bending stiffness (Figure 3.27), some of the differences are readily apparent and significantly affect behavior:

- Concrete conduit has a high bending stiffness and high hoop stiffness. Concrete is almost always classified as a rigid conduit (some efforts have been made to develop a thin wall concrete conduit, which can be semirigid).
- Corrugated and spiral rib steel and aluminum conduits have low bending stiffness and, as such, have been traditionally characterized

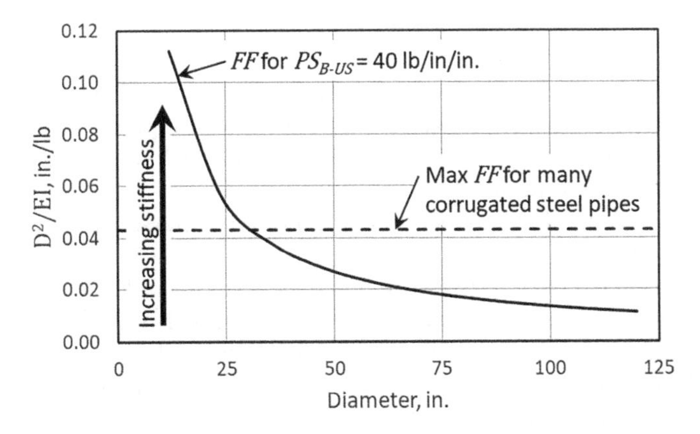

Figure 3.26 Flexibility factor versus diameter for a constant conduit stiffness, $PS_{B\text{-}US}$, of 40 lb/in./in.

Table 3.2 Typical conduit stiffness properties

	E_0	E_{50}	$\frac{E_0 I}{R^3}$	$\frac{E_{50} I}{R^3}$	$\frac{E_0 A}{R}$	$\frac{E_{50} A}{R}$
Conduit Type	*psi (MPa)*	*psi (MPa)*	*psi (MPa)*	*psi (MPa)*	*psi (MPa)*	*psi (MPa)*
Corrugated PE	110,000 (760)	22,000 (150)	4.4 (0.03)	0.9 (0.006)	2,200 (15)	450 (3.1)
Ribbed PVC	440,000 (3,000)	158,000 (1,100)	2.6 (0.018)	1.0 (0.007)	7,800 (54)	2,900 (20)
Corrugated Steel	29,000,000 (200,000)		15.5 (0.11)		123,000 (848)	
Reinforced Concrete	3,600,000 (25,000)		3,000 (21)		836,000 (5,700)	
Fiberglass	1,500,000 (10,000)	750,000 (5,200)	7.1 (0.05)	3.6 (0.02)	58,000 (400)	29,000 (200)

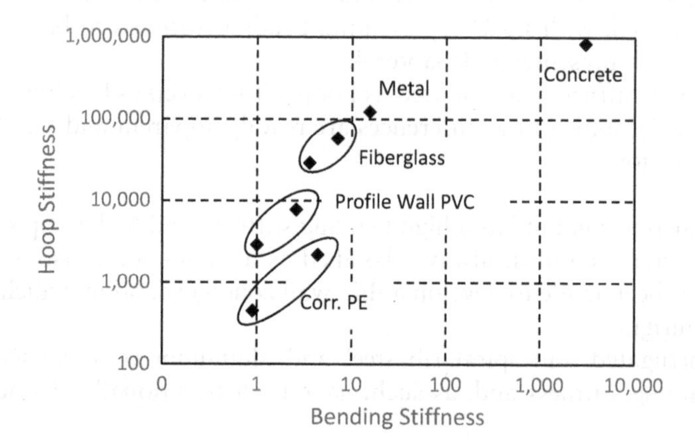

Figure 3.27 Relative hoop and bending stiffness of conduits.

as flexible conduits; however, these conduits have a high hoop stiffness, the significance of which is presented in Chapter 5. Although not shown, metal structures with deep corrugations likely classify as rigid or semirigid conduits.

- Fiberglass conduit has bending and hoop stiffnesses similar to metal conduits.
- Corrugated HDPE has a low bending stiffness and is classified as a flexible conduit, but the low long-term hoop stiffness changes the in-ground behavior. This feature was not considered in early design procedures for HDPE conduits since metal conduit theory was applied directly. Note that HDPE has the lowest short-term modulus and creeps by a factor of 5 over 50 years.

- PVC conduit stiffness is similar to corrugated HDPE but the higher long-term modulus and lower ratio of short to long-term modulus (approximately 3) keep the hoop stiffness notably higher than for the HDPE.

There are no specific boundaries separating flexible conduits from rigid conduits; rather, there is a transition where conduit behavior is semirigid (sometimes called semiflexible). Chapter 5 explores this transition and how it affects some designs.

3.2.2 Quality control tests

Both rigid and flexible conduit industries have adopted ring tests, in which a segment of conduit is loaded under diametrically opposed forces. These tests are convenient for testing but produce a more severe loading than a properly installed conduit would experience in the field. The relationship between test forces and in-ground forces in a conduit is used in simplified design methods, as presented in Chapter 2 and in greater detail in Chapter 7.

Flexible Conduit – Flexible conduits are often tested by loading between two parallel plates (Figure 3.25). Much of the US plastic conduit industry has adopted the parallel-plate test, ASTM D2412 *Standard Test Method for Determination of External Loading Characteristics of Plastic Pipe by Parallel-Plate Loading*, as a standard post-production test. Some international standards use a two-edge bearing test for flexible conduits. Equations (3.6) and (3.7) are accurate for relating load to deflection only at the start of the parallel-plate test. As the conduit geometry changes from a circle to an ellipse under load, the relationship becomes nonlinear as the top and bottom of the conduit flatten out against the loading plates. This nonlinearity is often ignored; however, DIN 16961-2, 2018, *Thermoplastics Pipes and Fittings with Profiled Wall and Smooth Pipe Inside – Part 2: Technical Delivery Specifications*, provides expressions to relate the applied force to deflection.

$$F = \frac{EI}{0.1488\, R_m^{\ 3}} \Delta Y\, C_1, \tag{3.9}$$

$$C_1 = \left(1 + \frac{0.2 \dfrac{\Delta Y}{D_m}}{0.1488}\right)^{-1}$$

where

F = Force applied to the parallel plates, lb, kN
ΔY = Vertical deflection, in., mm
C_1 = Correction factor based on deflection level

The C_1 correction factor reduces the applied force on the conduit by about 12% at a deflection of 10%.

The moment at the top and bottom of the conduit at the initiation of the parallel-plate test is

$$M = 0.3183\, W\, R_m. \tag{3.10}$$

Equations (3.7) and (3.8) can be combined with standard relationships for stress versus strain, the moment of inertia, and the section modulus to derive a direct relationship between relative displacement and width thickness ratio, t/D_m, for a solid wall conduit:

$$\varepsilon = 4.27 \frac{t}{D_m} \frac{\Delta Y}{D_m} \tag{3.11}$$

Equation (3.11) is sometimes used to relate the strain level of conduit in the ground to that in the parallel-plate test by replacing the constant with a shape factor, D_f, that can be larger or smaller than 4.27. This is discussed in Chapter 7.

Rigid Conduit – Rigid conduit industries use the three-edge bearing (TEB) test, ASTM C497 *Standard Test Methods for Concrete Pipe, Concrete Box Sections, Manhole Sections, or Tile*, as a standard quality control test (Figure 3.28). The TEB test uses two supports at close spacing under the conduit to prevent rolling, as the conduit is heavy and could pose risks to personnel.

The response to a TEB load, W_{TEB}, for an elastic ring with a uniform moment of inertia is as follows:

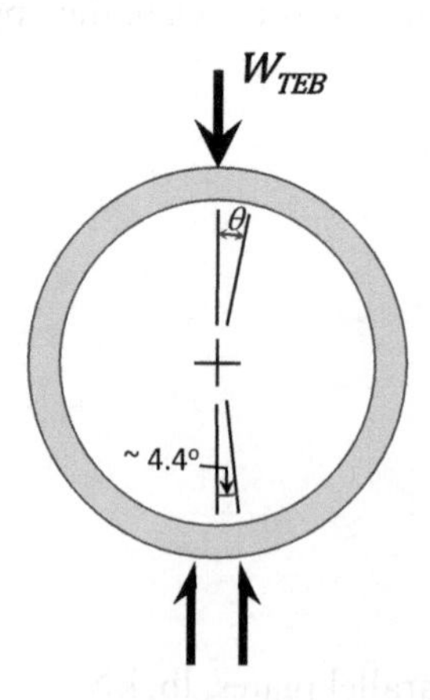

Figure 3.28 TEB test.

Moment:

$$M_{TEB} = W_{TEB}\, R_m \left(0.318 - 0.5 \sin\theta\right) = c_m\, W_{TEB}\, R_m \tag{3.12}$$

Shear:

$$V_{TEB} = -0.5\, W_{TEB} \cos\theta = c_v\, W_{TEB} \tag{3.13}$$

Thrust:

$$N_{TEB} = -0.5\, W_{TEB} \sin\theta = c_n\, W_{TEB} \tag{3.14}$$

Also important in the TEB test is the response to concrete conduit weight:

Moment:

$$M_P = W_p\, R\left(0160 - 0.081 \cos\theta - 0.159\,\theta \sin\theta\right) = c_m\, -_{Wp}\, W_P R \tag{3.15}$$

Shear:

$$V_P = W_p\left(-0.159\,\theta \cos\theta - 0.078 \sin\theta\right) = c_{v-Wp}\, W_p \tag{3.16}$$

Thrust:

$$N_P = W_p\left(-0.159\,\theta \sin\theta + 0.078 \cos\theta\right) = c_{n-Wp}\, W_p \tag{3.17}$$

where

M_{TEB} = Bending moment due to load, W_{TEB}, in TEB test, in.-lb/ft, kN-m/m
V_{TEB} = Shear force due to load, W_{TEB}, in TEB test, lb/ft, kN/m
N_{TEB} = Axial thrust force due to load, W_{TEB}, in TEB test, lb/ft, kN/m
W = Load in TEB test, lb/ft, kN/m
R_m = Mean radius of conduit, in., m
θ = Angle from crown, deg.
c_m, c_v, c_n = Coefficients for moment, thrust, and shear due to TEB load
M_P = Bending moment due to pipe weight, W_p, in TEB test, in.-lb/ft, kN-m/m
V_P = Shear force due to pipe weight, W_p, in TEB test, lb/ft, kN/m
N_P = Axial thrust force due to pipe weight, W_p, in TEB test, lb/ft, kN/m
W_p = Pipe weight, lb/ft, kN/m
$c_{m\text{-}wp}$, $c_{v\text{-}wp}$, $c_{n\text{-}wp}$ = Coefficients for moment, thrust, and shear due to pipe weight

Force coefficients at critical sections for the TEB test are presented in Table 3.3.

Table 3.3 Force coefficients for TEB test

	Applied Load, W_{TEB}			*Conduit Weight, W_p*		
	Moment	*Shear*	*Thrust*	*Moment*	*Shear*	*Thrust*
Location	C_m	C_v	C_N	$C_{m\text{-}Wp}$	$C_{v\text{-}Wp}$	$C_{N\text{-}Wp}$
Crown	0.318	−0.500	0.000	0.079	0.000	0.078
Springline	−0.182	0.000	−0.500	−0.090	−0.078	−0.250
Lower load point	0.280	0.499	−0.038	0.203	0.480	−0.115

Table 3.4 Crown and invert TEB moments (in.-lb/ft, kN-m/m)

	Diameter in. (mm)			
	24 (600)		*72 (1,800)*	
Location	*Crown*	*Invert*	*Crown*	*Invert*
Pipe Weight	251 (0.09)	647 (0.24)	5,147 (1.9)	13,250 (4.9)
TEB Load	11,448 (4.24)	10,067 (3.73)	103,032 (38.2)	90,604 (33.6)
Total	11,699 (4.34)	10,714 (3.97)	108,179 (40.1)	103,853 (38.5)

The coefficients in Table 3.3 are used to design conduits to pass the TEB test (Chapter 7).

The invert coefficients for the moment due to applied load suggest that the crown moment (C_m = 0.318) is greater than the lower load point coefficient (C_m = 0.280); however, the effect of the conduit weight, where the load point coefficient (C_m = 0.203) is much higher than the crown coefficient (C_m = 0.079) results in the two moments being within 10% of each other, as shown in Table 3.4. In practice, quality control staff typically monitor the conduit invert more closely than the crown.

REFERENCES

AASHTO (2020) *LRFD Bridge Design Specifications*, 9th Edition, American Association of Highway and Transportation Officials, Washington, DC.

AWWA (2014) Fiberglass Pipe Design, *Manual of Water Supply Practices—M45*, 3rd Edition, American Water Works Association, Denver, CO.

Heger, F.J., and Liepins, A.A. (1985) Stiffness of Flexurally Cracked Reinforced Concrete Pipe, *Proceedings ACI Journal*, Vol. 82(3), pp. 331–342.

Hsuan, Y.G., and McGrath, T.J. (1999) *NCHRP Report 429: HDPE Pipe: Recommended Material Specifications and Design Requirements*, Transportation Research Board of the National Academies, Washington, DC.

McGrath, T.J., Moore, I.D., and Hsuan, G.Y. (2009) *NCHRP Report 631: Updated Test and Design Methods for Thermoplastic Pipe*, Transportation Research Board of the National Academies, Washington, DC.

McGrath, T.J., Selig, E.T., and DiFrancesco, L.C. (1994) *Stiffness of HDPE Pipe, Buried Plastic Pipe Technology*, 2nd Volume, STP 1222, ASTM, Philadelphia, PA.

NCSPA (2018) *Corrugated Steel Pipe Design Manual*, 2nd Edition, National Corrugated Steel Pipe Association, Dallas, TX.

Uni-Bell (2012) *Handbook of PVC Pipe Design and Construction*, 5th Edition, Industrial Press Inc., New York, NY.

Chapter 4

Soils and soil models

Soil is a general term encompassing all the particulate natural materials on the earth's surface, including gravel, sand, silt, clay, and organic matter, as well as manufactured materials, such as crushed stone. In conduit design, soils must distribute loads around the conduit and provide support to limit its deformation (changes in shape). This chapter focuses on the properties of soil required to support conduits. It explores classification systems based on the particulate content to group soils with similar properties and introduces parameters and mathematical models that are needed to engineer conduit installations.

4.1 SOIL CLASSIFICATION SYSTEMS

Knowledge of the soil conditions into which a conduit will be installed and the backfills to be used as embedment allows an engineer to make decisions concerning support that will be provided to the conduit and the effort required to place and compact the backfill. Further, it allows engineering evaluations of possible disruptions of the backfill support, which might occur due to groundwater movements and weather events during the service life of the conduit.

Soil classification systems primarily use grain size (ASTM D422 *Standard Test Method for Particle-Size Analysis of Soils*) and Atterberg limits (ASTM D4318 *Standard Test Methods for Liquid Limit, Plastic Limit, and Plasticity Index of Soils*) to form groups with similar properties. The two primary classification systems used in the USA for conduit installations are AASHTO M145 *Standard Specification for Classification of Soils and Soil-Aggregate Mixtures for Highway Construction Purposes* and ASTM D2487 *Standard Practice for Classification of Soils for Engineering Purposes (Unified Soil Classification System)*. Grain size is almost always reported in SI units throughout the world. Thus, this section will treat SI as the primary unit system and inch-pound units as secondary. In the inch-pound system, it is common to designate sieve sizes by the number of openings per lineal inch of mesh. For example, a No. 4 sieve has four openings per lineal inch (16 openings per square inch) and will pass 4.75 mm (0.187 in.) particles.

DOI: 10.1201/9780429162619-4

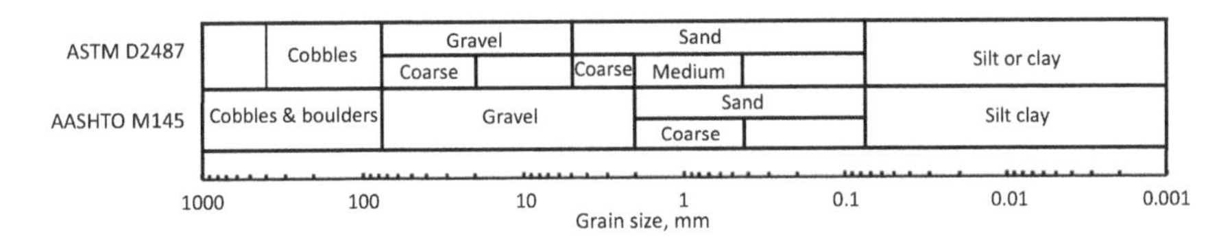

Figure 4.1 Grains size subcategories per AASHTO and ASTM classification systems.

Soil particles are first divided into two broad categories: coarse-grained and fine-grained (fines). Coarse-grained soils are retained on, and fines will pass through, a sieve with openings of 75 μm (0.0029 in., No. 200 sieve). Coarse-grained soils include sands and gravels and are the preferred embedment materials. Fines are made up of silts and clays. Soils containing significant quantities of fines are undesirable for embedment materials, as they can be difficult to compact and may not achieve the desired stiffness even after compaction. Cobbles (75–300 mm, 3–12 in.) and boulders (> 300 mm, 12 in.) are typically removed from a sample for classification purposes and are undesirable in the embedment zone. The AASHTO and ASTM standards vary on the size range and subdivisions of gravel and sand as shown in Figure 4.1.

Silt and clay particles can be a component of embedment materials, but the quantity is limited. Silts are distinguished from clays based on the liquid limit (LL), plastic limit (PL), and plasticity index (PI = LL – PL), determined in accordance with ASTM D4318 *Standard Test Methods for Liquid Limit, Plastic Limit, and Plasticity Index of Soils* and commonly known as the Atterberg limits. Silts typically have little or no strength when air-dry (non-plastic) and are susceptible to flowing when subjected to moving water. Clay particles are generally considered to be less than 0.002 mm (7.9×10^{-5} in.) in size and are cohesive.

Figure 4.2 shows the results of sieve analyses for several soils. For this test, a soil sample is placed on a screen with the largest opening of interest, which in turn is above a series of additional sieves with ever-decreasing opening sizes. The stack of sieves is then shaken so that all particles fall until they rest on a sieve with openings too small to pass through. The results are plotted on a semi-log plot as the percent finer by weight (percent passing) versus grain size. All the material that passes through any given sieve is the fraction of the total sample that is finer than that sieve opening size. Soils with uniform gradations (narrow range of grain sizes) are considered poorly graded (Soils 1, 4, and 5), while soils with a broader range of grain sizes are considered well graded (Soils 2 and 3) by ASTM, which is presented in Section 4.1.2. Soil 6 is a mixture of fine sand and silt but is not a suitable backfill for reasons discussed later.

4.1.1 AASHTO soil classification system

AASHTO M145 classifies soils using Table 4.1. The table is used by proceeding from left to right until a soil type meets all the requirements listed. Using Table 4.1 requires classifying the fines in a soil in accordance with the AASHTO plasticity chart (Figure 4.3).

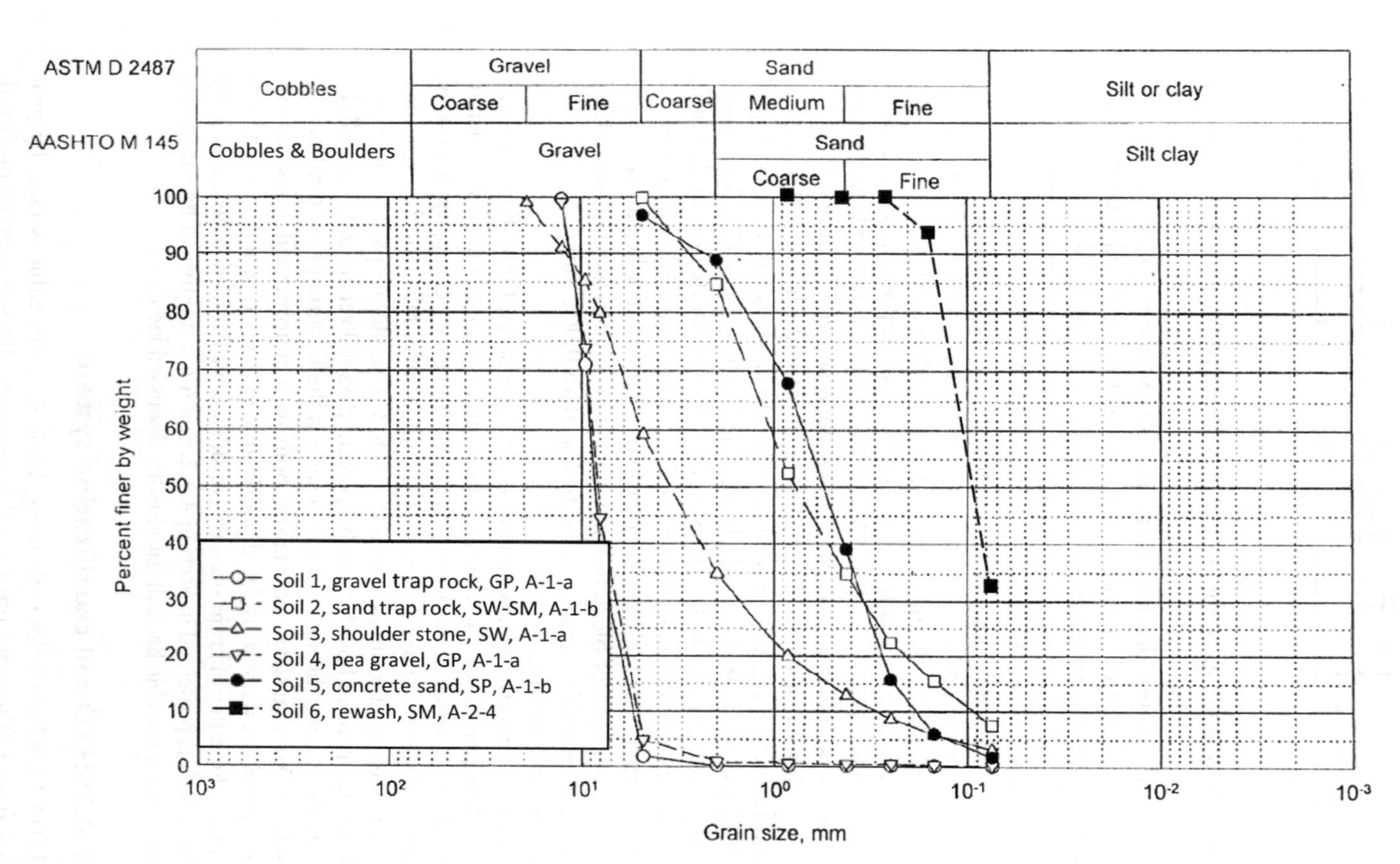

Figure 4.2 Sample grain-size analyses. (Adapted from McGrath et al. 1999)

Table 4.1 AASHTO soil classification table

General Classification	*Granular Materials (35% or Less Passing No. 200 Sieve (0.075 mm))*							*Silt-Clay MaterialsMore than 35% Passing No. 200 Sieve (0.075 mm)*			
Group Classification	*A-1*		*A-3*	*A-2*				*A-4*	*A-5*	*A-6*	*A-7*
	A-1-a	*A-1-b*		*A-2-4*	*A-2-5*	*A-2-6*	*A-2-7*				*A-7-5* *A-7-6*
(a) Sieve Analysis: percent passing											
(i) 2.00 mm (No. 10)	50 max										
(ii) 0.425 mm (No. 40)	30 max	50 max	51 min								
(iii) 0.075 mm (No. 200)	15 max	25 max	10 max	35 max	35 max	35 max	35 max	35 min	35 min	35 min	35 min
(b) Characteristics of fraction passing 0.425 mm (No. 40)											
(i) Liquid limit				40 max	41 min	40 max	41 min	40 max	41 min	40 max	41 min
(ii) Plasticity index	6 max		N.P.	10 max	10 max	11 min	11 min	10 max	10 max	11 min	11 min*
(c) Usual types of significant constituent materials	Stone Fragments Gravel and Sand		Fine Sand	Silty or Clayey Gravel Sand				Silty Soils		Clayey Soils	
(d) General rating as subgrade	Excellent to Good							Fair to Poor			

Source: From AASHTO M145, published by the American Association of State Highway and Transportation Officials, Washington, DC. Used with permission.

* If the plasticity index is equal to or less than (liquid limit – 30), the soil is A-7-5 (i.e., PL > 30%) If the plasticity index is greater than (liquid limit – 30), the soil is A-7-6 (i.e., PL < 30%)

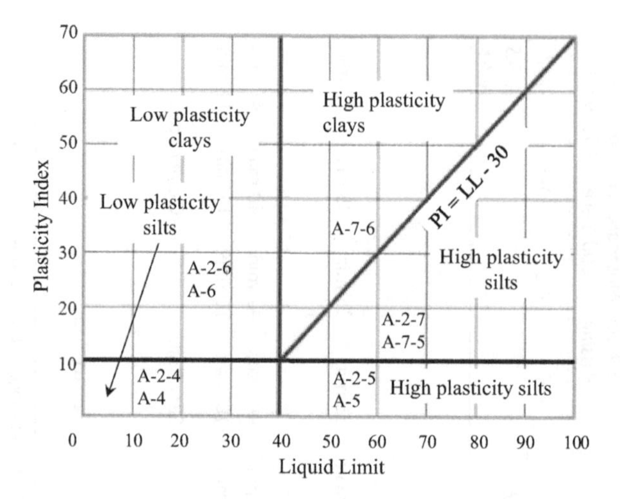

Figure 4.3 AASHTO plasticity chart. (From AASHTO M145, published by the American Association of State Highway and Transportation Officials, Washington, DC. Used with permission)

Key points of the AASHTO system include the following:

- Granular materials limit the fines content to 35% passing. Soils with more than 35% fines are considered silt clays.
- A-1 and A3 soils limit the fines content to 15% and 25% or less, respectively, but also limit the PI of the fines to 6, which eliminates clay from these soil classifications.
- A-2 soils allow up to 35% fines, more than A-1or A-3 soils. The classifications A-2-4 to A-2-7 follow a progression of increased plasticity, i.e., low plasticity silts, low plasticity clays, high plasticity silts, high plasticity clays.
- A-4 through A-7 soils have no upper limits on fines content and follow the same progression of increased plasticity as the A-2-4 soils.

AASHTO soil classifications suitable for use as conduit embedment are discussed in Section 4.1.4.

4.1.2 ASTM soil classification system

ASTM D2487 classifies soils as coarse-grained when the fines content is 50% or less and fine-grained if the fines content is greater than 50%. This differs from AASHTO, which limits fines content to 35% for granular materials. (Table 4.1). In ASTM, a gravel has more gravel than sand particles, and a sand has more sand than gravel particles. Silts (M) and clays (C) have greater than 50% passing the 0.075 mm (No. 200) sieve and are distinguished based on the Atterberg limits. Organic soils (O) are not considered suitable for the embedment zone and are not discussed further.

ASTM D2487 provides a flow chart for classifying coarse-grained soils (Figure 4.4). Sands and gravels with less than 5% fines are either well graded (GW, SW) or poorly graded (GP, SP) based on the coefficients of uniformity (C_u) and curvature (C_c) as calculated per ASTM D2487. Soils with 5%–12% fines are given dual designations, such as GW-GM or SP-SC, where the M and C identify the nature of the fines. Soils with 12%–50% fines are given two-letter designations, such as GC or SM, based on the classification of the fines. The flow chart extends the classification beyond the group symbol based on relative sand and gravel content with group names, such as "well-graded gravel with silt and sand."

A similar flow chart (Figure 4.5) identifies the classification process for fine-grained soils. Fines are classified as low (L) or high (H) plasticity, giving group names, such as ML or CH, per the ASTM plasticity chart (Figure 4.6). The ASTM chart varies from the AASHTO plasticity chart in the boundary between low and high plasticity and the location of the diagonal line separating clays from silts. Some silt and clay materials can be used for conduit embedment, but they require careful control of moisture content and significantly increased effort during placement and compaction relative to coarse-grained materials.

Field Identification of Soil – ASTM D2488 *Standard Practice for Description and Identification of Soils (Visual-Manual Procedures)* provides guidelines that allow field personnel to readily identify soil classifications in the field with a few simple supplies. It is useful to become familiar with these procedures in the event of uncertainty on an active project. The practice is applicable to both backfill materials and in situ soil.

4.1.3 International Standards Organization (ISO) soil classification

ISO Standard 14688-1 *Geotechnical Investigation and Testing – Identification and Classification of Soils – Part 1: Identification and Description* uses different boundaries to distinguish the various grain-size ranges, as shown in Figure 4.7. ISO Standard 14688-2 *Geotechnical Investigation and Testing – Identification and Classification of Soils – Part 2: Principles for a Classification* classifies soil using the coefficients of uniformity and curvature (C_u, C_c) for coarse-grained soils and the soil plasticity (LL, PL, PI) for fine-grained soils. While this standard presents soil classifications in slightly different forms, those differences are not significant for the purpose of selecting embedment soils for conduit backfill. ISO 14688 does not provide designations for combinations of soil particles similar to the ASTM and AASHTO groups noted earlier.

4.1.4 Soil groups used for conduit embedment

Specifying backfill gradations and properties for conduit embedment can be difficult, as there are two requirements that can be in conflict. Backfill

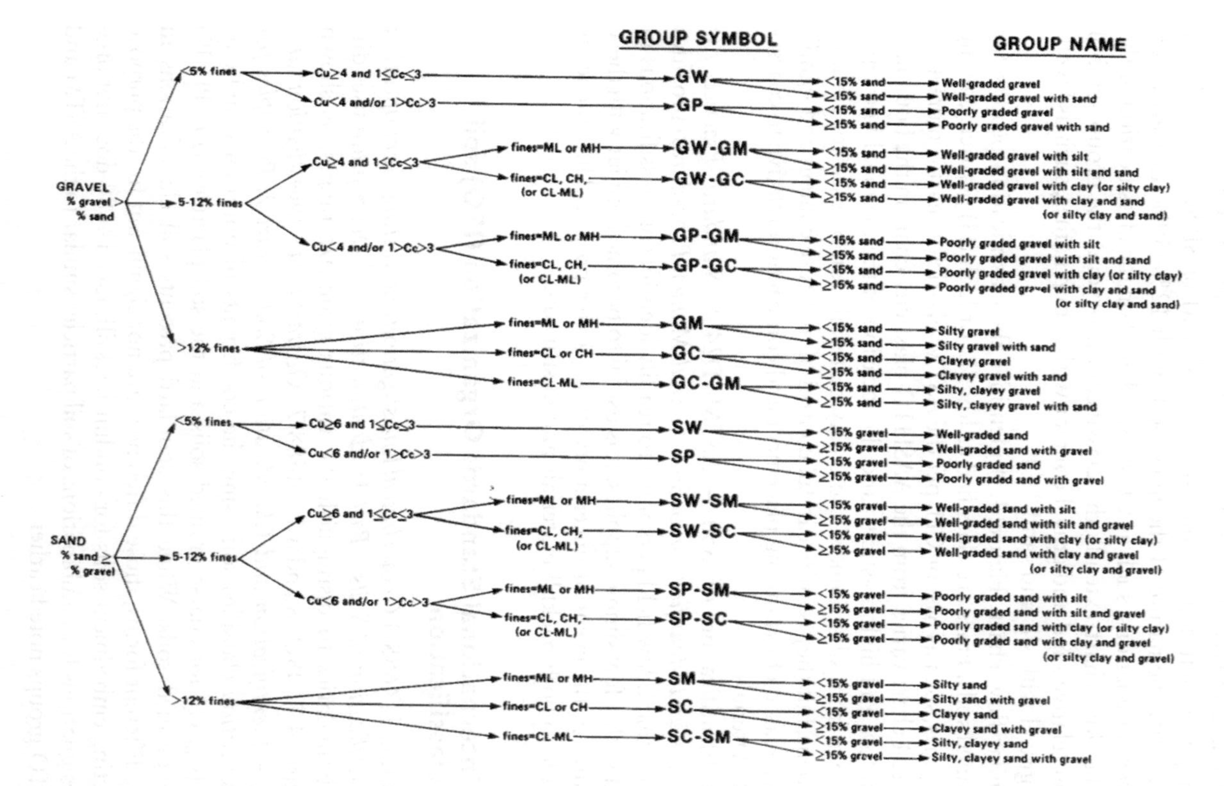

Figure 4.4 ASTM flow chart for classifying coarse-grained soils. (ASTM D2487 reprinted with permission. Copyright ASTM International, www.astm.org)

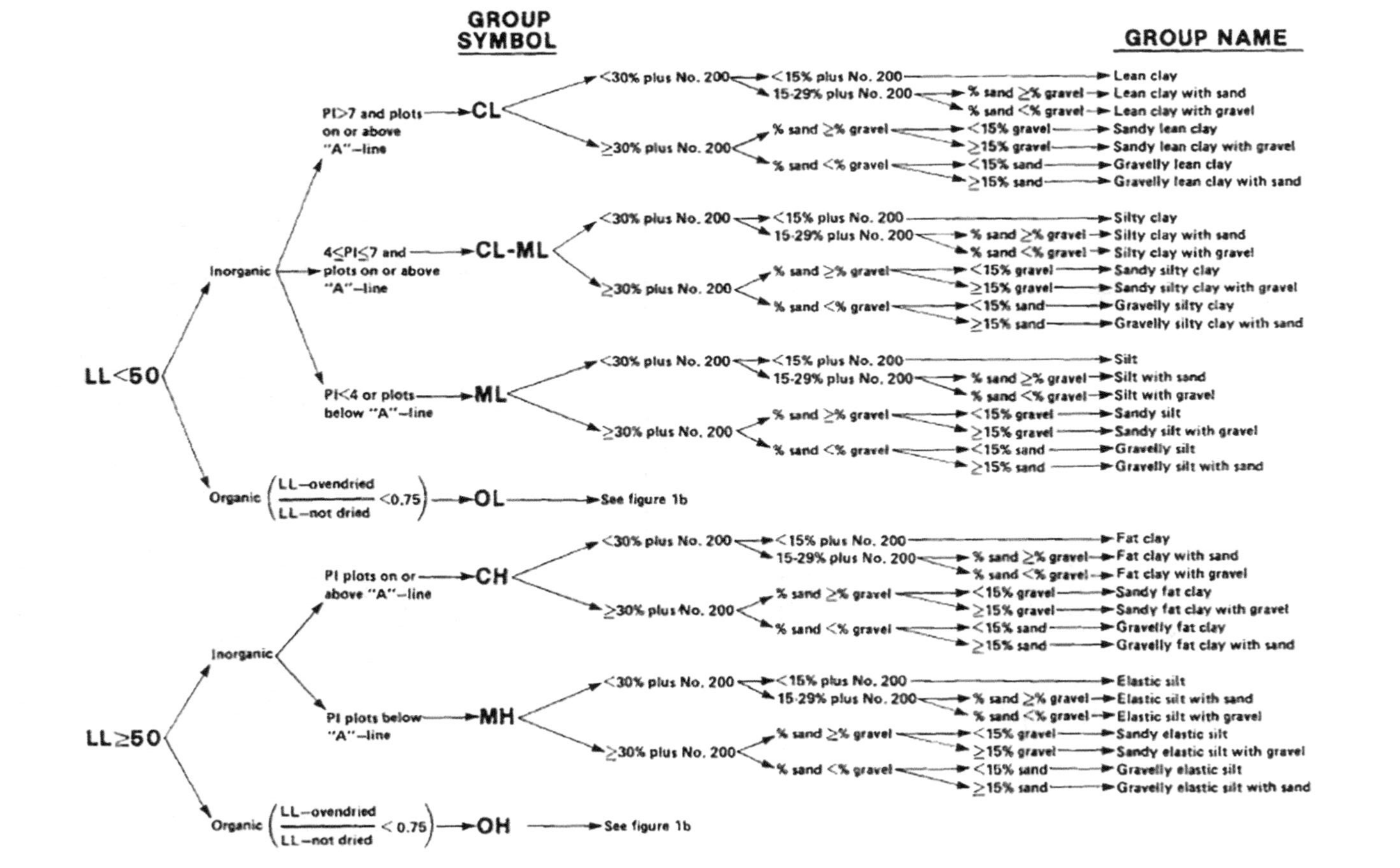

Figure 4.5 ASTM flow chart for classifying fine-grained soils. (ASTM D2487 reprinted with permission. Copyright ASTM International, www.astm.org)

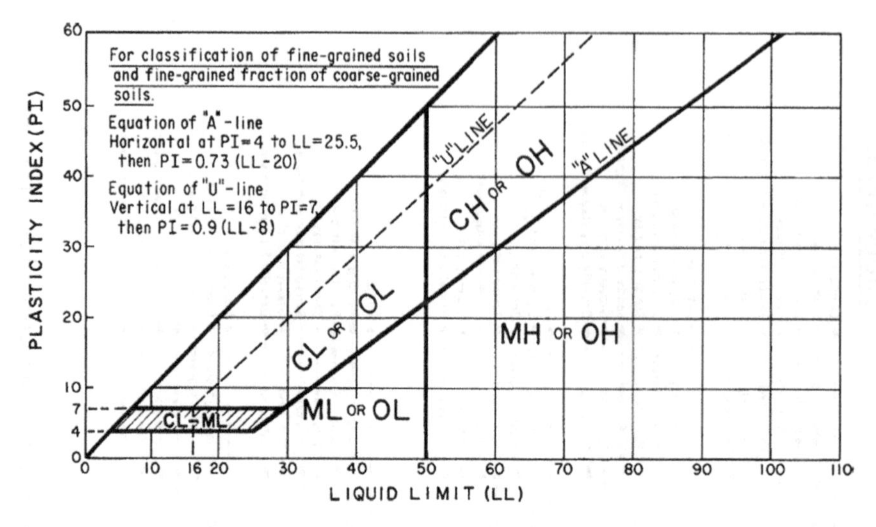

Figure 4.6 ASTM plasticity chart. (Reprinted with permission from ASTM D2487. Copyright ASTM International, www.astm.org)

ISO 14688-1	Boulders; Cobbles; Gravel (Coarse, Medium, Fine); Sand (Coarse, Medium, Fine); Silt (Coarse, Medium, Fine); Clay
ASTM D2487	Cobbles; Gravel (Coarse, Fine); Sand (Coarse, Medium, Fine); Silt or clay
AASHTO M145	Cobbles & Boulders; Gravel; Sand (Coarse, Fine); Silt clay
Grain Size, mm	1000 100 10 1 0.1 0.01 0.001

Figure 4.7 Comparison of ASTM, AASHTO, and ISO grain-size classification.

specifications should be restrictive enough to ensure a backfill material can provide the necessary support to the conduit but broad enough to allow the use of available materials. A broad specification allows contractors to take advantage of the savings possible by purchasing local soils but might include some unacceptable gradations while a narrow specification may assure good performance but require costly backfill. As discussed in this section and in Section 8.3, ASTM, AASHTO, and many other standards provide broad specifications. Specifiers and field engineers should require the submittal of backfill gradation curves for review and approval prior to construction.

The AASHTO LRFD Bridge Design Specifications (AASHTO LRFD, 2020) Article 12.4.1.3 provides guidance on soil groups from AASHTO M145 and ASTM D2487 that are considered suitable for embedment of conduits. These are distinguished based on the type of structure (Table 4.2).

Article 27.5.2.2 of the AASHTO LRFD Bridge Construction Specifications (AASHTO Construction 2017) and ACPA 1998 describe embedment materials for concrete conduits that are grouped into categories that provide similar performance (Table 4.3).

Table 4.2 Allowable embedment soils per AASHTO LRFD bridge specifications

Structure	*AASHTO soil groups**	*ASTM soil groups*
Standard flexible conduits and concrete structures	A-1, A-2, A-3	GW, GP, SW, SP, GM, SM, SC, GC
Metal box culverts and long-span structures with cover less than 12 ft	A-1, A-2-4, A-2-5, A-3	GW, GP, SW, SP, GM, SM
Long-span metal structures with cover more than 12.0 ft	A-1, A-3	GW, GP, SW, SP, GM, SM, SC, GC
Structural plate structures with deep corrugations	A-1, A-2-4, A-2-5, A-3	GW, GP, SW, SP, GM, SM, SC, GC
Thermoplastic, fiberglass, and steel-reinforced thermoplastic culverts	A-1, A-2-4, A-2-5, A-3 $\leq$ 50% passing 0.150 mm (0.0059 in., No. 100) sieve and $\leq$ 20% passing No. 200 (0.0029 in., 0.075 mm) sieve	

* Metal conduit manufacturers limit the sand content in some backfills similar to the thermoplastic guidelines, but the limits are not included in AASHTO specifications.

Table 4.3 Concrete conduit embedment materials

	Allowable Soil Types	
Soil Group	*ASTM D2487*	*AASHTO M145*
Category I	SW, SP, GW, GP	A-1, A-3
Category II	GM, SM, ML Also GC, SC with less than 20% fines	A-2, A-4
Category III	GL, MH, GC, SC	A-5, A-6

Tables 4.2 and 4.3 list the broad soil groups capable of providing the desired support to a buried conduit, but there are differences between the ASTM and AASHTO classifications. Engineers should understand the differences and make a determination if additional constraints are required to provide suitable embedment for a specific project.

- AASHTO soil A-3 is a fine sand, with a minimum of 51% required to pass a 425 μm (No. 40) sieve and a maximum of 10% fines. A sand backfill is generally considered a good embedment; however, in the extreme, an A-3 soil could consist of 100% passing a 150 μm (No. 100) sieve and 90% retained on a 75 μm (No. 200) sieve, which is essentially silt. Such a backfill is difficult to compact and could easily flow in the presence of moving water and is thus undesirable as an embedment soil. To avoid this, the specifications for thermoplastic conduits restrict the fine sand content by requiring a minimum of 50% to be retained on a 150 μm (No.100) sieve. Note that an ASTM SP soil could raise the same concern. One study (McGrath et al. 1999) attempted to use a silty sand (SM) for bedding and backfill; however, when the bedding was compacted in the afternoon, it was saturated by a minimal amount of groundwater overnight, leaving it in an unacceptable running soil condition. This required changing the backfill to an SP sand with 60% retained on a 425 μm (No. 40) sieve.

- All AASHTO soils that are acceptable for embedment of flexible conduits are limited to 35% fines. This compares to acceptable ASTM soils (e.g., SM), which allow fines content of up to 50%.
- Further, ASTM uses dual names for coarse-grained soils with 5%–12% fines– for example, SP-SM for a sand with silty fines in that range. Since SP and SM can both be acceptable backfills, it follows that dual-classification soils would be acceptable, yet AASHTO specifications provide no guidance.

Section 4.3 and Chapter 8 present and discuss backfill groups based on comparable stiffness and compactibility.

4.2 UNIT WEIGHT (DENSITY) TESTING

The desirable soil property for limiting conduit deformations and stresses is soil stiffness. While there are some methods available to measure soil stiffness directly, it is most common to correlate the soil type and soil dry unit weight or density to determine the stiffness properties. This section discusses the standard tests for laboratory testing to determine the density that can be achieved under controlled conditions.

The most common laboratory test is commonly called the standard Proctor test: ASTM D698 *Standard Test Methods for Laboratory Compaction Characteristics of Soil Using Standard Effort (12,400 ft-lbf/ft³ [600 kN-m/m³])* and AASHTO Standard T99 *Standard Method of Test for Moisture-Density Relations of Soils Using a 2.5-kg (5.5-lb) Rammer and a 305-mm (12-in.) Drop*. In this test, a soil sample is placed in a 4-in. (101.6 mm) diameter cylindrical mold in three layers, with each layer tamped 25 blows by a 5.5 lb (24.5 N) hammer dropped from a height of 12 in. (305 mm). The total energy imparted to the soil is part of the ASTM Standard title. There is also a modified Proctor test: ASTM D1557 *Standard Test Methods for Laboratory Compaction Characteristics of Soil Using Modified Effort (56,000 ft-lbf/ft³ [2,700 kN-m/m³])* and AASHTO T180 *Moisture-Density Relations of Soils Using a 4.54 KG (10 LB) Rammer and a 457 MM (18 IN.) Drop*. This test applies 4.5 the energy to the soil relative to the standard effort test by using a heavier hammer, a greater drop height, and five layers. The modified Proctor test is not in common use for conduit backfill. Both standards allow the use of a 6-in. (152.4 mm) mold, which requires increasing the number of blows per layer to 56. Use of the larger mold is advised for some coarse-grained soils with large particles. The Proctor test, as developed by Ralph Proctor, used a weight that was forcibly pushed down, and the energy applied was thus operator dependent. The ASTM and AASHTO tests use a dropped weight for better uniformity and therefore do not use "Proctor" in the standard titles. This text continues the common practice of calling both "Proctor tests" despite the difference in the test procedure.

The Proctor test is conducted by preparing the soil sample to a moisture content expected to be less than the content to achieve maximum dry unit weight. The test is then conducted at that moisture content and at successively higher moisture content until the dry unit weight achieved is lower than the previous test. The reported result of the test is the maximum dry unit weight and the moisture content at which the maximum dry unit weight is achieved, called the "optimum" moisture content. Up to the optimum water content, adding moisture improves the movement of particles over each other, achieving a greater dry unit weight. However, as the water content approaches saturation, the water interferes with the soil particles moving to denser configurations. In the field, the installation contractor will achieve the highest compaction result with the least compactive effort if the soil is kept near the optimum moisture content.

Figure 4.8 presents standard Proctor test results for the soils in Figure 4.2. The figure shows the data points and the "saturation line" indicating 100% water content, in this case for an assumed specific gravity, G, of 2.9. There is some scatter in the data points, but Soils 2, 3, 5, and 6, all of which have some fine sand and silt content, show the desired curve rising to a maximum

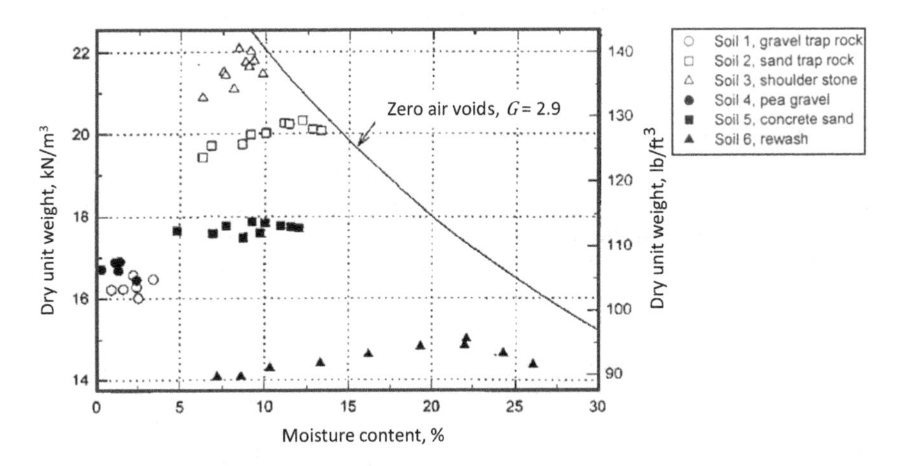

Figure 4.8 Standard Proctor test results. (McGrath et al. 1999)

and then falling off. Soils 1 and 3 show a cluster of points and an inability to hold water above a small percentage. This demonstrates the applicability of the Proctor test to soils that can hold water distributed throughout the specimen. If this is not the case, water added to the specimen quickly drains out, and the classic moisture-density curve yielding a maximum dry unit weight and optimum moisture content cannot be obtained.

When the Proctor test does not achieve the classic moisture-density relationship, the relative density tests, ASTM D4253 *Standard Test Methods for Maximum Index Density and Unit Weight of Soils Using a Vibratory Table* or ASTM D7382 *Standard Test Methods for Determination of Maximum Dry Unit Weight of Granular Soils Using a Vibrating Hammer* may be used. ASTM D4253 and ASTM D 4254 *Standard Test Methods for Minimum Index Density and Unit Weight of Soils and Calculation of Relative Density* are used in tandem to determine the 0% and 100% densities for use in calculating relative density. ASTM D4254 determines a minimum dry unit weight by placing soil in a mold through a funnel with a minimal drop height, while ASTM D4253 determines a maximum dry unit weight by placing a dry soil specimen in a mold and vibrating it.

Relative density is not a good method to evaluate field density. Figure 4.9 compares the results of Proctor tests, relative density tests, and the dumped density at optimum moisture for the same six soils considered in Figure 4.2. Observations from this figure include the following:

- The maximum relative density varies from 94% to 107% of the maximum standard Proctor density, which is judged to be reasonable agreement.
- The density, when dumped at optimum moisture content, is much less than the minimum relative density. This is because wet soils with fines will tend to bulk together, leaving air voids that will not occur when the soil is placed dry in accordance with the minimum relative density test. The fewer fines in the soil, the closer the two loose densities become.

The vibration table to conduct the maximum relative density test is not available in many areas. If this is the case, Figure 4.9 suggests that using the average maximum density from the Proctor tests is a reasonable approach to setting a maximum field density requirement. This would apply to Soils 1 and 4 in Figure 4.8. The average of several tests provides a reasonable maximum, even though the classic moisture-density relationship is not achieved.

Relative density should not be used as a method of field monitoring, as the density range between minimum and maximum relative density can be small. For example, Soil 1, gravel trap rock, has maximum and minimum relative densities of 103 and 87 lb/ft^3 (16.1 and 13.4 kN/m^3). Thus a 5% change in relative density is 0.8 lb/ft^3 (0.14 kN/m^3), which means small

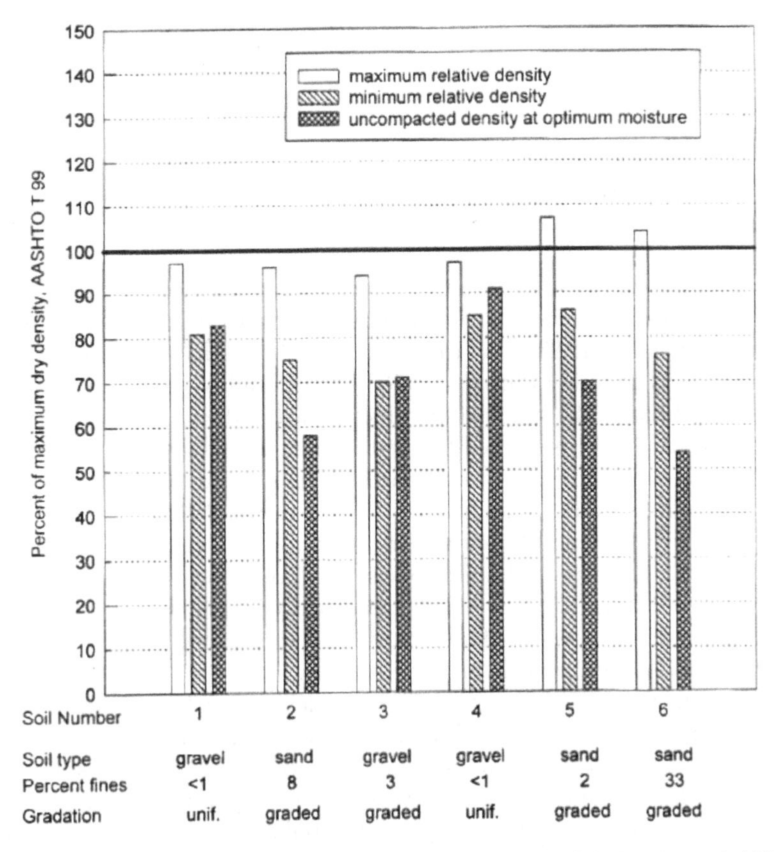

Figure 4.9 Loose and compacted density of backfill soils. (McGrath et al. 1999)

variations in laboratory tests or field test results could mean failing soils that have been compacted to acceptable densities. This concern is amplified in ASTM D4253, which notes, "There are published data to indicate that these test methods have a high degree of variability." This subject is examined again in Section 9.3. Calculating field density as a percentage of the maximum density from ASTM D4253 or D7382, or even from a Proctor test, avoids this problem.

4.3 STIFFNESS VERSUS UNIT WEIGHT

The desired soil property for conduit design is stiffness, as represented by a modulus of elasticity. Since the most common test to evaluate soil placement is unit weight (density), the relationship between stiffness and unit weight is important. As noted in Chapter 2, Marston and Spangler both recognized the importance of soil stiffness for rigid and flexible conduits. Soil models used for conduit design vary from simple linear parameters for use in the Iowa formula to multivariable, nonlinear, stress-dependent models used in finite element analyses. The evolution of soil models for conduit design over several decades includes the following:

- The Iowa formula developed by Spangler used the modulus of passive resistance to represent soil stiffness with the units of psi/in. of displacement (kPa/mm of displacement), which is not a true soil property as the modulus value varies with the size of the loaded area, like the modulus of subgrade reaction used in bearing capacity calculations. Watkins and Spangler (1958) took the product of the modulus of passive resistance and conduit radius and called it the modulus of soil resistance, E', which became the standard parameter used in the Iowa formula.
- Howard (1977) presented the first widely accepted table of *E'* values for use in the Spangler formula and later updated his original recommendations based on additional data (Howard, 2015).
- The development of finite element analysis opened the possibility of more sophisticated soil models. The computer program CANDE (Katona et al., 1976), developed specifically for the purpose of analyzing conduits, at first included linear and stress-dependent models and later added nonlinear hyperbolic models.
- Duncan et al. (1980) developed a model with a hyperbolic Young's modulus and a power law bulk modulus for use in analyzing conduits. Duncan also provided input parameters for several types of soils.
- Selig (1988) built on Duncan's work, using a hyperbolic Young's modulus and proposed a new hyperbolic bulk modulus. Selig proposed properties for several soils for use in finite element analysis. Current design procedures for conduits in the AASHTO LRFD Bridge Design Specifications for concrete, plastic, and some metal conduits are based on Selig's properties.
- McGrath (1998) proposed that, for the purposes of conduit design, *E'* and the constrained modulus, M_s, could be considered the same parameter and, using the Selig model, developed M_s values based on soil type and stress level for use in the Iowa formula.

There are many finite element programs currently available that are suitable for conduit analysis and many suitable soil models for use in those programs. This text will focus on the models noted earlier because of their common use in conduit design practice and because the principles discussed are widely applicable to other programs and models.

4.3.1 Modulus of soil reaction, *E'*

When Spangler developed the Iowa formula, the concept of designing flexible conduits based on the stiffness of the embedment soils was new. Marston and his colleagues had incorporated soil stiffness into the design method for rigid conduits but had not developed a design model for flexible conduits. Little was known about the variations in soil stiffness based on soil type and gradation at that time. Spangler conducted three experiments that included six installation conditions, five with a lean clay backfill and one with a "pit run gravel." Spangler reported back-calculated values for the modulus of passive resistance, which convert to modulus of soil reaction, *E'*, values of about 300 psi (2 MPa) and 600 psi (4 MPa) for tamped and untamped backfill. Interestingly, the *E'* value for the test with untamped pit run gravel was greater than the value from the tests with tamped clay backfills. Spangler included the higher value from this test with the average for the tamped backfills but did not specifically note the extra stiffness in the gravel backfill, missing an opportunity to glean additional information from his test data.

Howard (1977) surveyed the deflections in many conduit installations and proposed a table of *E'* values based on compaction levels and soil types that came into common usage in the Iowa formula (Table 4.4).

Howard's *E'* table was a significant step forward in understanding the support provided to conduits by different soil types at different densities. Table 4.4 shows the following:

- The soil groups in the table are constructed to consider every possible classification in ASTM D2487. This is an improvement over the AASHTO use of ASTM soil groups as backfill classifications, which, as discussed earlier, do not include all soils (e.g., soils with 5%–12% fines or dual-classification soils).
- The table shows a difference in behavior in fine-grained soils if the coarse-grained content is more or less than 30%. This is a boundary in ASTM D2487. A CL soil with less than 30% coarse-grained particles is a lean clay, while one with more than 30% is called a sandy or gravelly clay. The same boundary is included for silts.
- Values of *E'* = 1,000 psi (7 MPa) form a diagonal in the table, as it is achieved with dumped crushed rock but requires increased compaction with increasing fines content, reaching 95% for fine-grained soils.
- Only crushed stone provides an *E'* value of 1,000 psi (7 MPa) or greater in the dumped condition.
- Howard limits the application of the *E'* values to depths of fill up to 50 ft.
- The line "accuracy in terms of percent deflection" shows values of 2% for dumped and slight compaction down to 0.5% for high compaction. This emphasizes the improved predictability of deflection levels as construction quality and control are increased. Further, the increased variability in the mean indicates a higher standard deviation along the length of a conduit. Since a conduit will fail a deflection test due to the maximum deflection, not the mean, controlling the variability is important. This is discussed in more detail in Chapter 7.

Table 4.4 Original Howard *E'* table, psi (MPa)

Embedment Classification (ASTM D2487)	*Degree of Compaction of Embedment*			
	Dumped	*Sligh < 85% SP < 40% RD*	*Moderate ≥ 85% to < 95% SP ≥ 40% to < 70% RD*	*High ≥ 95% SP ≥ 70% RD*
Crushed rock Not more than 25% passing 3/8 in. (9.5 mm) sieve and not more than 12% fines; maximum size not to exceed 3 in. (75 mm)	1,000 (7)	3,000 (21)		
Clean coarse-grained soils Sands or gravels with 12% or less fines GW, GP, SW, SP, or soils beginning with one of these symbols (i.e., GP-GM)	200 (1.5)	1,000 (7)	2,000 (14)	3,000 (21)
Sandy or gravelly fine-grained soils Soils with medium to no plasticity with 30% or more coarse-grained articles CL, ML (or CL-ML, CL/ML, ML/CL) Coarse-grained soils with fines Sands, gravels with more than 12% fines GC, GM, SC, SM, or any soil beginning with one of these symbols (i.e., SC/CL)	100 (0.7)	400 (3)	1,000 (7)	2,000 (14)
Fine-grained soils Soils with medium to no plasticity with less than 30% coarse-grained articles CL, ML (or CL-ML, CL/ML, ML/CL)	50 (0.3)	200 (1.5)	400 (3)	1,000 (7)
Highly compressible fine-grained soils CH, MH, OH, OL, or any soils containing one of those symbols (i.e., CL/CH)	No data available; consult a competent soil engineer or use E' = 0 NOT RECOMMENDED			
Accuracy in terms of percent deflection	±2%	±2%	±1%	±0.5%

Notes: SP = standard Proctor density, RD = relative density. Values are applicable only for cover depth of 50 ft or less.

Source: (Howard, 1977, with permission from ASCE)

Table 4.5 Revised Howard *E'* table, psi (MPa)

Soil Group	*Uncompacted*	*Compacted** *Moderate 85% or 90%*	*High 95% or 100%*
Class I Crushed rock	1,000 (6.9)	6,000 (41)	
Class II GW, GP, SW, SP	500 (3.4)	2,000 (14)	4,000 (28)
Class III GC GM SC SM sCL, sML gCL gML	200 0.14	1,000 (6.9)	2,500 (17)
Class IV CL ML	100 0.69	400 (2.8)	1,500 (10)
Class V CH MH OH OL Pt	Not Recommended for Embedment		

* Percent of Maximum standard Proctor density
Source: Howard, 2015, with permission.

Howard's updated table of *E'* values in his book *Pipeline Installation 2.0* (Howard, 2015) is presented in Table 4.5. With more data available, the table shows higher values for soils with high compaction levels and much higher values for compacted crushed rock.

Differences from Howard's original table include the following:

- The *E'* values for highly compacted backfill all increased.
- The dumped and slightly compacted columns on the original table have been replaced with a single column labeled uncompacted with mostly intermediate values of *E'* relative to Table 4.4.
- The soil descriptions appear much simpler but upon reading the notes associated with the revised table, the soil groups are the same in Tables 4.4 and 4.5. The lowercase "s" and "g" refer to sandy and gravelly soils, which are defined in Appendix X5 in ASTM D2487.
- A qualifier is added for SP soils requiring greater than 50% retained on a 425 µm (0.0165 in., No. 40) sieve. This is consistent with the AASHTO limit on fine sands for thermoplastic conduit backfill, as noted earlier.
- The limitation to conduits with less than 50 ft depth of cover is continued.

Conduit deflection in trench installations may be influenced by the stiffness of the backfill and the soil in the trench wall. A method for addressing this in design is presented in Section 7.7.1.

The revised Howard values for *E'* (Table 4.5) are recommended if *E'* is the preferred design parameter for soil stiffness. The values in Table 4.4 are used in the discussion below because some points can be seen more clearly. An alternative design parameter for deflection prediction is the constrained modulus, M_s, introduced in the following section.

4.3.2 Nonlinear soil models

The advent of computerized finite element methods allowed the analysis of buried conduits with much more sophisticated models for soil behavior. The US Federal Highway Administration saw this possibility and funded the development of CANDE (Katona et al. 1976), a finite element program for the analysis of conduits and, in particular, long-span metal conduits that were being developed at that time. CANDE is an acronym for Culvert ANalysis and DEsign. Shortly after this, Duncan et al. (1980) proposed a nonlinear soil model based on a hyperbolic Young's modulus and a power law bulk modulus. Duncan also proposed parameters to determine properties for several types of soils.

Subsequently, Selig (1988) built on Duncan's work by developing a hyperbolic bulk modulus soil model for use with Duncan's Yong's modulus model. The equations forming the basis of the Duncan/Selig model include the following:

$$(\sigma_1 - \sigma_3) = \frac{\varepsilon_v}{\dfrac{1}{E_i} + \dfrac{\varepsilon_v}{(\sigma_1 - \sigma_3)_u}}, \tag{4. 1}$$

where

σ_1 = Major principal stress, psi, kPa
σ_3 = Minor principal stress, psi, kPa
$(\sigma_1 - \sigma_3)$ = Deviator stress, psi, kPa
ε_v = Vertical strain, in./in., mm/mm
E_i = Initial Young's modulus = $K\ P_a\ (\sigma_3/P_a)^n$, psi, kPa
$(\sigma_1 - \sigma_3)_u$ = Ultimate deviator stress, psi, kPa
K and n = Model parameters
P_a = Atmospheric pressure, used to nondimensionalize K and n

$$\sigma_m = \frac{B_i\ \varepsilon_{vol}}{1 - \dfrac{\varepsilon_{vol}}{\varepsilon_u}}, \tag{4.2}$$

where

σ_m = mean stress = $(\sigma_1 + 2\ \sigma_3)/3$, psi, kPa

B_i = initial bulk modulus, a model parameter
ε_{vol} = volumetric strain
ε_u = ultimate volumetric strain, a model parameter

The ultimate deviator stress is determined by the failure ratio, R_f, and the deviator stress at failure $(\sigma_{3\text{-}1}-\sigma_3)_f$:

$$R_f = \frac{(\sigma_1 - \sigma_3)_f}{(\sigma_1 - \sigma_3)_u}, \tag{4.3}$$

where

$$(\sigma_1 - \sigma_3)_f = \frac{2\,C\cos\phi + 2\,\sigma_3\sin\phi}{1 - \sin\phi} \tag{4.4}$$

C = cohesion intercept, psi, kPa
ϕ = friction angle, degrees

The friction angle is a measure of soil strength used in geotechnical engineering. Soils fail in shear, but the strength varies with the state of the stress in the soil. As the confining stresses in the soil increase, so does the shear capacity. For example, in a uniform soil mass, the vertical and horizontal stresses increase with depth, which increases the shear strength.

As the failure envelope is often curved, the friction angle is determined as

$$\phi = \phi_o - \Delta\phi \log_{10}\left(\frac{\sigma_3}{P_a}\right), \tag{4.5}$$

ϕ_o = friction angle for $\sigma_3 = P_a$,
$\Delta\,\phi$ = decrease in friction angle for a ten-fold increase in σ_3.

The second parameter required for analysis is the bulk modulus, determined as

$$B = B_i\left[1 + \frac{\sigma_m}{(B_i\,\varepsilon_u)}\right]^2. \tag{4.6}$$

Thus, the Selig model requires eight parameters: K, n, C, ϕ, ϕ_o, $\Delta\phi$, and R_f to determine the tangent Young's modulus and B_i and ε_u to determine the bulk modulus. The tangent modulus is the slope of the soil stress-strain curve at any given point. This gives the instantaneous stiffness of soil, which is required for finite element analysis where conduit models are loaded in a step-by-step manner, including both soil and vehicular loads. Simplified

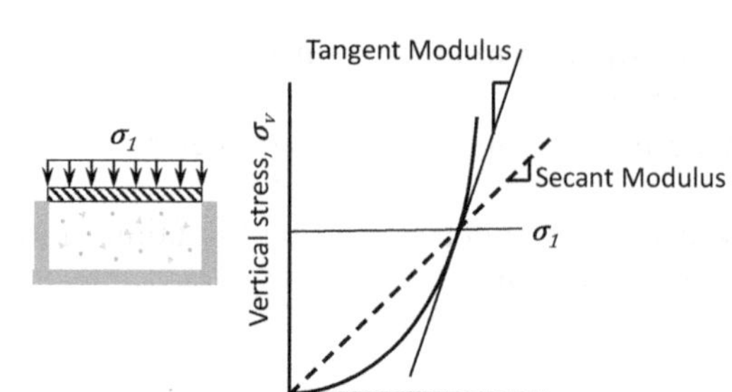

Figure 4.10 Constrained and secant moduli in a one-dimensional compression test.

design uses the secant modulus, which is the slope of a line from the origin of the stress-strain curve to a specific point on the curve. This provides the average soil stiffness over the expected range of loading, allowing a single-step calculation. The tangent and secant moduli are illustrated in Figure 4.10 for the stress-strain curve in a one-dimensional compression test.

Selig developed parameters for three soil types, which he called SW, ML, and CL, based on the soils he tested. The naming is somewhat unfortunate, as the three soils are used in practice to represent much broader groups of soils. For concrete conduits, AASHTO LRFD uses the SW, ML, and CL labels, but AASHTO Construction has relabeled the groups Category I, Category II, and Category III. For thermoplastic conduits, AASHTO LRFD calls the three soil groups Sn, Si, and Cl, which are the labels used in Table 4.6, presenting the Selig parameters.

The soils tested to determine the Selig parameters are described in Table 4.7. All three soils are conservative relative to many possible soils that would fit into the Sn, Si, and Cl categories.

Items of interest shown by this table include the following:

- The gravelly sand classification and the optimum moisture content of 7.4% suggest that the Sn soil contains a fair amount of fine sand and or silt. Thus, the properties should be conservative relative to clean sands and gravels.
- The sandy silt and A4 classifications suggest that the Si soil has more than 35% fines and less than 15% gravel. Note that A-4 soils are not allowed according to the AASHTO specifications, although the relatively low liquid limit and plasticity index would qualify this material as an A-2-4 soil except for the fines content.
- The silty clay and A-6 classifications also suggest that the Cl soil had a fines content greater than 36%, but similar to the Sn soil, the relatively low liquid limit and plasticity index would suit an A-2-6 soil. This soil would not meet any requirements for acceptable conduit backfill in AASHTO.

Table 4.6 Parameters for Selig soil model

Soil Group %SP	*K*	*n*	*C* psi, kPa	ϕ_o deg.	$\Delta\phi$ deg.	R_f	B_i/P_a	ε_u
	Tangent Young's Modulus Parameters						*Bulk Modulus Parameters*	
Sn100	1,300	0.90	0.0	54	15	0.65	108.8	0.01
Sn95	950	0.60	0.0	48	8.0	0.72	74.8	0.02
Sn90	640	0.43	0.0	42	4.0	0.75	40.8	0.05
Sn85	450	0.35	0.0	38	2.0	0.80	12.7	0.08
Sn80	320	0.35	0.0	36	1.0	0.90	6.1	0.11
Si95	440	0.40	4.0	34	0.0	0.95	48.3	0.06
Si90	200	0.26	3.5	32	0.0	0.89	18.4	0.10
Si85	110	0.25	3.0	30	0.0	0.85	9.5	0.14
Si80	75	0.25	2.5	28	0.0	0.80	5.1	0.19
Si50	16	0.95	0.0	23	0.0	0.55	1.3	0.43
Cl95	120	0.45	9.0	15	4.0	1.00	21.2	0.13
Cl90	75	0.54	7.0	17	7.0	0.94	10.2	0.17
Cl85	50	0.60	6.0	18	8.0	0.90	5.2	0.21
Cl80	35	0.66	5.0	19	8.5	0.87	3.5	0.25

Note: %SP refers to the numerical suffix attached to the soil group, which indicates the percent of maximum standard Proctor dry density.

Source: Selig, 1988, with permission from ASCE.

Table 4.7 Soils tested to derive Selig properties

Label	*Soil Type*	*Classification* D 2487	*Classification* AASHTO	*Consistency (%)* Liquid Limit	*Consistency (%)* Plasticity Index	*D 698 Compaction* Maximum Dry Density (pcf, kN/m^3)	*D 698 Compaction* Optimum Moisture (%)
Sn	Gravelly sand	SW	A-1	–	NP	138 (23.2)	7.4
Si	Sandy silt	ML	A-4	20	4	119 (18.7)	12.1
Cl	Silty clay	CL	A-6	32	15	103 (16.2)	21.0

Source: Selig, 1988, with permission from ASCE.

These parameter properties proposed by Selig were used in the finite element analyses for concrete conduit that led to the standard installations now incorporated into AASHTO. This model has also been incorporated into CANDE.

4.3.3 Constrained (one-dimensional) modulus

Several researchers observed that the modulus of soil reaction should be close in value to the one-dimensional (constrained) soil modulus, M_s, a parameter within elasticity theory representing uniaxial compression (Figure 4.10).

Since E' represents the average stress-strain behavior while loading a soil, the secant modulus is of most interest. Krizek et al. (1971) reported that several researchers had studied the relationship between E' and M_s and concluded there is no single relationship between the two parameters, with M_s varying from 0.7 to 1.5 E_s. McGrath (1998) revisited this subject using the Selig model to compute secant values of M_s at various stress levels. Table 4.8 presents the M_s values proposed by McGrath with the in-lb units in ksi for consistency with AASHTO LRFD. These values capture the increase in stiffness that occurs as the confining soil stresses increase.

McGrath compared the M_s values in Table 4.7 with Howard's (1977) E' values and values of stress-dependent E' proposed by Hartley and Duncan (1987) and concluded the equality of $E' = M_s$ is reasonable for the purpose of estimating the deflection of buried conduits when using Spangler's Iowa formula for design.

Table 4.8 Secant M_s values for Selig soil parameters

a. In.-lb Units

Stress Level (psi)	*Soil Type and Compaction Condition (ksi)*									
	Sn100	*Sn95*	*Sn90*	*Sn85*	*Si95*	*Si90*	*Si85*	*Cl95*	*Cl90*	*Cl85*
1	2.35	2.00	1.28	0.47	1.42	0.67	0.36	0.53	0.26	0.13
5	3.45	2.60	1.50	0.52	1.67	0.74	0.39	0.63	0.32	0.18
10	4.20	3.00	1.63	0.57	1.77	0.75	0.40	0.69	0.36	0.20
20	5.50	3.45	1.80	0.65	1.88	0.79	0.43	0.74	0.40	0.23
40	7.50	4.25	2.10	0.83	2.09	0.90	0.51	0.82	0.46	0.29
60	9.30	5.00	2.50	1.00	2.31	1.03	0.60	0.90	0.53	0.35

b. SI Units

Stress Level (kPa)	*Soil Type and Compaction Condition (MPa)*									
	Sn100	*Sn95*	*Sn90*	*Sn85*	*Si95*	*Si90*	*Si85*	*Cl95*	*Cl90*	*Cl85*
7	16.2	13.8	8.8	3.2	9.8	4.6	2.5	3.7	1.8	0.9
35	23.8	17.9	10.3	3.6	11.5	5.1	2.7	4.3	2.2	1.2
69	28.9	20.7	11.2	3.9	12.2	5.2	2.8	4.8	2.4	1.4
138	37.9	23.8	12.4	4.5	13.0	5.4	3.0	5.1	2.7	1.6
275	51.7	29.3	14.5	5.7	14.4	6.2	3.5	5.6	3.2	2.0
413	64.1	34.5	17.2	6.9	15.9	7.1	4.1	6.2	3.6	2.4

Source: McGrath, 1998, with permission from ASCE.

Gemperline (2010) conducted a series of one-dimensional compression tests on several crushed rock materials using a special apparatus to accommodate the large particle sizes. The test results suggested M_s values of 2 to 4 ksi (14 to 28 MPa) for loosely placed stone and up to 5 to 10 ksi (35 to 70 MPa) for tamped stone.

4.3.4 Soil groups corresponding to M_s parameters

Section 4.1.4 and Tables 4.2 and 4.3 noted the difficulties in developing groups of soils that provide comparable stiffness in conduit installations and M_s or *E'* design values. Howard (2015) has matched his *E'* table with soils classified in accordance with ASTM D2487. It has been a challenge to similarly align soils classified in accordance with AASHTO M145, mostly because of the differences between the two groups. If one accepts that deflections calculated with the Spangler formula are approximate at best, then the approach should be to keep the soil groups in a form that is readily incorporated into specifications. Table 4.9 proposes such an approach.

4.3.5 Compactability

Compactability is a term used to describe the ease with which a soil may be compacted and, in turn, the ease with which required soil stiffness can be achieved. Selig (1988) conducted standard Proctor density tests with variable energy input as part of his study to develop the soil model discussed earlier. This consisted of reducing the number of blows per layer to achieve energy levels of approximately 0%, 17%, and 50% of the standard test energy to go along with the complete test (Figure 4.11). The figure shows that SW (Sn in Table 4.8) soil reached 90% of the maximum standard Proctor density with only 17% of the normal Proctor test energy, while the CL (Cl in Table 4.8) soil required 50% of the normal Proctor energy to reach 90% of maximum density.

Combining Selig's energy data with Howard's original *E'* table provides a plot of the energy required to achieve stiffness for these three soils. Figure 4.12 demonstrates that soils with increasing amounts of fines require increased compactive effort to achieve the same percent of maximum Standard Proctor density but achieve less stiffness than soils with limited fines.

Figure 4.12 shows that the clean coarse-grained soil can be compacted to 95% of maximum standard Proctor density with only 30% of the total energy applied in a Proctor test but provides an *E'* = 3,000 psi (21MPa). In comparison, the fine-grained soil requires twice the energy to achieve only 1/3 of the stiffness of the clean coarse-grained soil.

Table 4.9 Soil groups for conduit embedment

ASTM Soil Groups[a]	*AASHTO Soil Group*[b]	*Group Names*[c] *(Concrete/ Thermoplastic/ Fiberglass)*	M_s *Group*[g]
Crushed rock[d]: 100% passing 1-1/2 in. (37 mm) sieve, ≤ 15% passing No. 4 (4.75 mm) sieve, ≤ 25% passing 3/8 in. (9.5 mm) sieve and ≤ 12% passing No. 200 (0.075 mm) sieve	...	– Class I SC1	
Clean coarse-grained soils[e]: SW, SP, GW, GP, or any soil beginning with one of these symbols with ≤ 12% passing No. 200 (0.075 mm) sieve[c]	A1, A3[f]	Category I Class II SC2	Sn
Coarse-grained soils with fines: GM, GC, SM, SC, or any soil beginning with one of these symbols, containing > 12% passing No. 200 (0.075 mm) sieveSandy or gravelly fine-grained soils: ML, CL, or any soil beginning with this symbol, with ≥ 30% retained on the No. 200 (0.075 mm) sieve	A-2-4, A-2-5	Category II Class III SC3	Si
Fine-grained soils: CL or ML with < 30% retained on No. 200 (0.075 mm) sieve or any soil beginning with one of these symbols	A-4, A-2-6, A-2-7	Category III Class IV SC4	Cl
MH, CH, OL, OH, PT	A5, A6, A7	– Class V SC5 Not for use as embedment	

[a] Based on ASTM D2487 with some additions and modifications to improve suitability as embedment materials.
[b] Based on AASHTO M145.
[c] Group names for concrete, thermoplastic, and fiberglass conduits are from AASHTO Construction Article 27, ASTM D2321and AWWA Manual M45, respectively. Metal conduit design practice does not use specific group names.
[d] Class I materials are labeled crushed rock, but rounded materials such as pea gravel can also provide high stiffness with minimal compaction.
[e] Materials such as broken coral, shells, and recycled concrete, with ≤ 12% passing a No. 200 (0.75 mm) sieve, are considered Class II materials. These materials should only be used when approved by the engineer, as there may be additional considerations concerning long-term suitability.
[f] A-3 soils are restricted to a maximum of 50%, passing a No. 100 (0.15 mm) sieve. Restrictions should be considered on the portion passing the No. 100 sieve to avoid a material overly sensitive to moisture.
[g] M_s for crushed rock is often taken as Sn values at 95% and 100% maximum standard Proctor density for the dumped and compacted conditions. Howard (2015) suggests specific E' values for crushed rock.

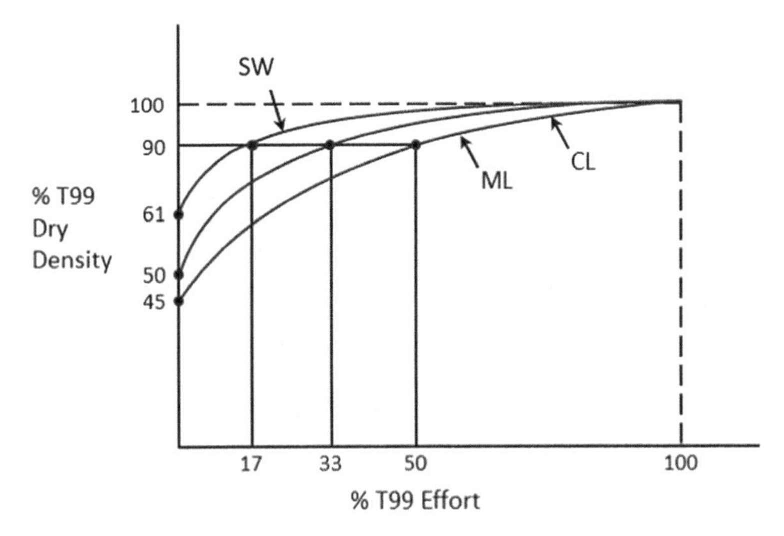

Figure 4.11 Effect of soil type and compaction effort on density. (Selig, 1988, with permission from ASCE)

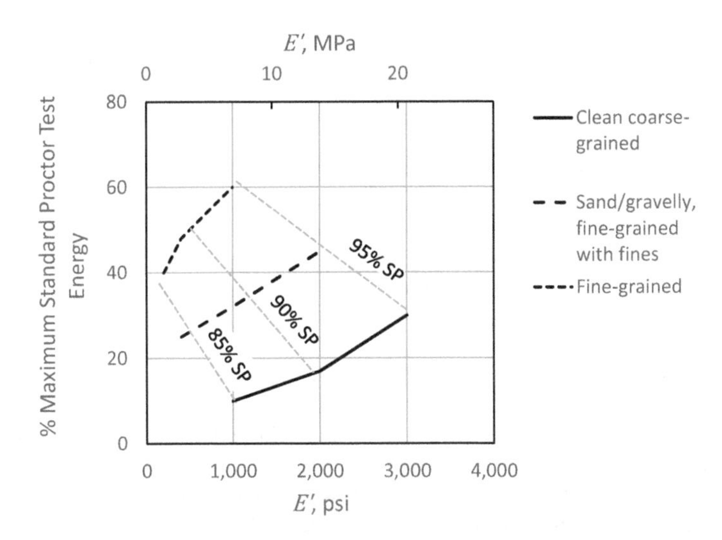

Figure 4.12 Energy to achieve soil stiffness.

4.4 CONTROLLED LOW-STRENGTH MATERIAL (FLOWABLE FILL)

Controlled low-strength material (CLSM) or flowable fill consists of cement, sand, water, and fly ash or other additives to attain a flowable consistency. In some applications, native soils can be used to replace the sand. The goal for conduit installations is to provide backfill that flows into the trench and fills the haunches without manual labor, reaches the stiffness required to support the conduit, and will still be excavatable if repairs are required. Very low strength meets the stiffness requirement, and very low strength is needed to meet the excavatable requirement. Howard (2015) suggests 40–80 psi as a suitable strength. McGrath and Hoopes (1997) investigated the use of air entrainment to improve flowability and to reduce long-term strength gain to maintain excavatability.

REFERENCES

AASHTO (2017) *AASHTO LRFD Bridge Construction Specifications*, 4th Edition, AASHTO, Washington, DC.

AASHTO (2020) *AASHTO LRFD Bridge Design Specifications*, 9th Edition, AASHTO, Washington, DC.

Duncan, J.M., Byrne, P., Wong, K.S., and Mabry, P. (1980) *Strength, Stress-Strain and Bulk Modulus Parameters for Finite Element Analyses of Stresses and Movements in Soil Masses, Department of Civil Engineering Report No. UCB/GT/80-01*, University of California, CA.

Gemperline, M., Gemperline, E., and Gemperline, C. (2010) *Large Scale Constrained Modulus Test*, Final Report to the Plastic Pipes Institute, Irving, TX.

Hartley, J.P., and Duncan, J.M. (1987) E' and Its Variations with Depth, *Journal of Transportation Engineering ASCE*, Vol. 113, No. 5, pp. 538–553.

Howard, A.K. (1977) Modulus of Soil Reaction Values for Buried Flexible Pipe, *Journal of Geotechnical Engineering, American Society of Civil Engineers*, Vol. 103, No. GT-1.

Howard, A.K. (2015) *Pipeline Installation 2.0*, Relativity Publishing, Lakewood, CO.

Katona, M.G., Smith, J.M., Odello, R.S., and Allgood, J.R. (1976) *A Modern Approach for the Structural Design and Analysis of Buried Culverts, FHWA Report No. FHWA-RD-77-5*, Federal Highway Administration, Washington, DC.

Krizek, R.J., Parmelee, R.A., Kay, J.N., and Elnaggar, H.A. (1971) *Structural Analysis and Design of Culverts, NCHRP Report 116*, National Cooperative Highway Research Program, Highway Research Board, National Academy of Sciences, Washington, DC.

McGrath, T.J. (1998) Replacing E' with the Constrained Modulus in Buried Pipe Design, *Proceedings, Pipelines in the Constructed Environment*, J.P. Castronovo and J.A. Clark, Eds., American Society of Civil Engineers, Reston, VA.

McGrath, T.J., and Hoopes, R.J. (1997) Bedding Factors and E' Values for Buried Pipe Installations Backfilled with Air-Modified CLSM, *The Design and Application of Controlled Low-Strength Materials (Flowable Fill), ASTM STP 1331*, A.K. Howard and J.L. Hitch, Eds., American Society for Testing and Materials W., Conshohocken, PA.

McGrath, T.J., Selig, E.T., Webb, M.C., and Zoladz, G.V. (1999) *Pipe Interaction with the Backfill Envelope, FHWA-Rd-98-191*, Federal Highway Administration, Washington, DC.

Selig, E.T. (1988) Soil Parameters for Design of Buried Pipelines, *Pipeline Infrastructure – Proceedings of the Conference*, American Society of Civil Engineers, New York, NY, pp. 99–116.

Watkins, R.K., and Spangler, M.G. (1958) Some Characteristics of the Modulus of Passive Resistance of Soil, a Study in Similitude, *Proceedings HRB*, Vol. 37, pp. 576–583.

Chapter 5

Soil-conduit interaction

Buried conduits interact with the surrounding soil. This interaction results in both "load" on the conduit and "support" to the conduit, although the difference between load and support is not always clear. This chapter expands on the ideas presented in Chapter 2 and explores key features of soil-conduit interaction that should be understood to design buried conduits effectively. Starting with simple elastic models to demonstrate how all buried conduit behavior fits into a continuum based on basic conduit and soil parameters, the chapter then expands to consider actual field conditions used to design buried piping systems and describes how some of these behaviors are addressed in current practice.

5.1 BEHAVIOR OF ELASTIC SYSTEMS

Prior to developing finite element methods of analysis, simple elastic models were developed to understand the behavior of horizontal tubes buried in elastic soil masses. Burns and Richard (1964) developed a closed-form elastic solution for this condition that can be used to investigate conduit response as well as soil stresses. The model for the Burns and Richard (Burns) solution is shown in Figure 5.1.

With the assumption of a weightless, infinite, isotropic, homogenous elastic medium, the model is loaded with a vertical soil pressure, p_v, based on the unit weight of the soil, γ_s, and depth of fill over the top of the conduit, H, and a lateral pressure, K_o ($\gamma_s H$), where K_o is the assumed coefficient of lateral earth pressure at rest. The conduit was modeled as a thin ring with assigned values for modulus of elasticity, E, area, A, and moment of inertia, I. Solutions were derived for two conditions: a bonded conduit-soil interface (no-slip) and a frictionless interface (full-slip). Solutions for the conduit were derived for moment, M, thrust, N, and radial displacement, w. Solutions for the soil include the radial and tangential pressure at the soil-conduit interface as well as soil stresses throughout the elastic medium. Burns selected the constrained modulus, M_s, as the appropriate parameter to model soil behavior. The model assumes the soil around the conduit to be a

DOI: 10.1201/9780429162619-5

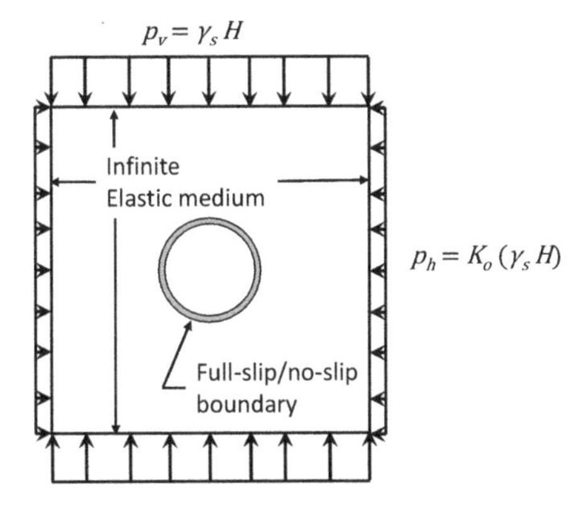

Figure 5.1 Burns and Richard model for soil-conduit interaction.

uniform continuum; thus, the bedding and haunch zones are all fully filled and have a stiffness identical to the rest of the soil mass. This condition is rarely met in field installations, yet the model still has considerable value in expanding our understanding of soil-conduit interaction.

The Burns model is incorporated into the computer program CANDE (Katona et al. 1976, CANDE 2022) as a Level 1 solution. In CANDE, Burns solutions can be implemented in a stepwise fashion to model nonlinear soil behavior, but still only with a uniform soil mass. Finite element programs can be used to complete an analysis using the same assumptions as the Burns solution.

5.1.1 Interaction parameters

Chapter 3 introduced the key conduit stiffness parameters, and Chapter 4 introduced soil behavior. The key parameters for soil-conduit interaction are the ratios of soil stiffness to conduit stiffness:

Bending stiffness ratio:

$$S_B = \frac{M_s}{PS_B} = \frac{M_s\, R_m{}^3}{E\, I}, \text{ and} \tag{5.1}$$

Hoop stiffness ratio

$$S_H = \frac{M_s}{PS_H} = \frac{M_s\, R_m}{E\, A}. \tag{5.2}$$

Note that the bending stiffness uses R_m to represent the conduit size rather than D_m. This is to keep with the derivation of the Burns and Richard solution. For corrugated or other nonprismatic conduit walls, R_m and D_m represent the radius and diameter of the centroid of the conduit wall cross-section.

Also, note that the stiffness ratios increase with increasing soil stiffness and decrease with increasing conduit stiffnesses.

Other relationships used in the solution include the assumed coefficient of lateral earth pressure at rest, *K*:

$$K_o = \frac{\nu}{1-\nu} \tag{5.3}$$

and the relationship between Young's modulus and constrained modulus:

$$M_s = \frac{E_s(1-\nu)}{(1+\nu)(1-2\nu)}, \tag{5.4}$$

where

M_s = Constrained (one-dimensional) modulus of the soil, psi, MPa
E_s = Young's modulus of the soil, psi, MPa
ν = Poisson's ratio of the soil

Poisson's ratio is the ratio of lateral expansion strain to vertical compression strain under a vertical stress with no lateral restraint. Since lateral movement in a soil mass in the field is restricted, a horizontal stress will develop when vertical stress is applied. If the soil is fully restricted from any lateral spreading, the ratio of horizontal to vertical stress is K_o, the at-rest lateral pressure ratio.

Poisson's ratio of soil varies widely if all possible soil types and installation conditions are considered; however, if the soil types are limited to those acceptable as conduit backfill, a value of ν = 0.3 may be used for most designs. Figure 5.2 demonstrates this with a plot of *VAF* versus S_H

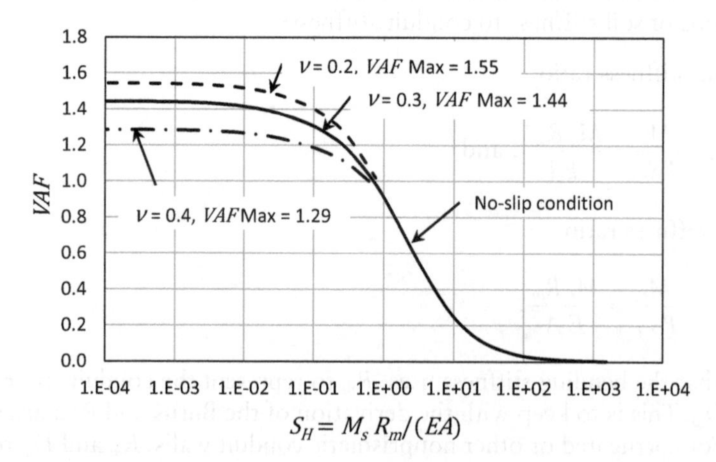

Figure 5.2 Effect of Poisson's ratio on vertical arching factor.

and $\nu = 0.2$, 0.3, and 0.4 based on the Burns and Richard. The figure shows that for conduits with high hoop stiffness, values of Poisson's ratio greater than 0.3 produce a lower vertical arching factor and, thus, a reduced load on the conduit, i.e., a conservative solution. Poisson's ratios of less than 0.3 produce higher arching factors and a greater load on the conduit. At a value of $\nu = 0.2$, the predicted *VAF* is 7% higher than for $\nu = 0.3$. This error is judged acceptable for routine work given the approximate nature of conduit backfill design.

All conduit response figures plot with a similar shape as shown in Figure 5.2. The curves are made up of three regions. The first region is a horizontal portion where the conduit response is insensitive to the soil stiffness, $S_H > 100$ in Figure 5.2, where *VAF* goes to zero because the soil stiffness is so high or the conduit stiffness is so low that the conduit carries no load. The second portion of the curve is a sloping transition where the soil stiffness and conduit stiffness both contribute to the conduit response. The response in this region is often referred to as semirigid or semiflexible behavior, depending on one's preference. The third portion of the curve is a second horizontal portion where the conduit response is again insensitive to the soil stiffness, $S_H < 0.01$ in Figure 5.2, and the *VAF* remains constant.

5.1.2 Conduit response to soil load

5.1.2.1 Hoop compression

Conduit and soil stiffness parameters both contribute to the calculation of S_H and S_B, and thus each type of conduit will only correspond to a limited range of values on plots such as Figure 5.2. Figure 5.3 shows the curve for *VAF* versus S_H for the Burns no-slip and full-slip conditions. Calculations were conducted to determine the *VAF* for the concrete, metal (corrugated steel), and plastic (corrugated polyethylene) conduits listed in Table 5.1. Note that the curve marked "metal" does not strictly follow the typical curve shape. In this instance, and some others, the interaction of S_B and S_H creates an extra wave in the curve, but the basic S-curve (solid line) is always conservative for determining the VAF. Figure 5.3 shows the range of *VAF* for soil stiffnesses, M_s, between 500 and 10,000 psi (3.5 to 70 MPa) for each of the three types of conduit. The *VAF* range for plastic conduit is shown for both the short and long-term modulus of elasticity, which demonstrates the reductions in *VAF* that can occur after backfilling is complete due to stress relaxation of the polymer.

Figure 5.3 demonstrates the design considerations for hoop compression in each conduit type:

- The concrete conduit has the smallest S_H, i.e., the highest conduit hoop stiffness, and is in the range where the *VAF* is essentially constant at about 1.4 for the no-slip condition. This means the load experienced by

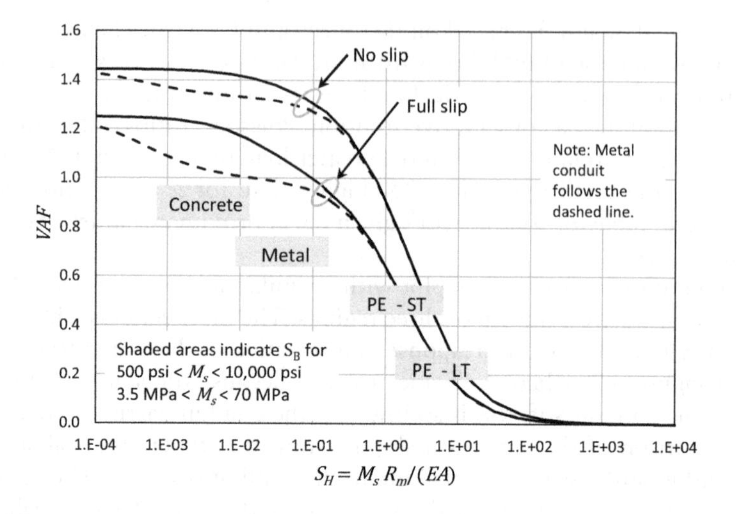

Figure 5.3 VAF vs. hoop stiffness, S_H.

the conduit is essentially independent of soil stiffness. Current design theory for concrete conduits in embankment installations is based on analysis with a bonded interface and uses *VAF* of 1.35 to 1.45.

- The metal conduit has a moderately higher value of S_H relative to concrete but is still predominantly in the range where *VAF* is constant. These conduits have long been designed assuming *VAF* = 1.0, which is less than the value of 1.25 for the flat portion of the full-slip condition; however, as noted in Chapter 7, metal conduits are designed with a load factor of 2 on thrust, which is adequate to accommodate the actual load condition.
- The corrugated PE conduit plots on the sloping portion of the curves in Figure 5.3, where the VAF is dependent on soil stiffness. The design for hoop thrust is based on the long-term condition. The long-term earth load on these conduits can be much less than the soil prism load because of the low modulus of elasticity and, hence, the low hoop stiffness. Like metal conduits, plastic conduits have traditionally been designed with a *VAF* of 1.0 and a load factor of 2 on thrust, which covers the condition of *VAF* = 1.2.

Thus, in resisting hoop compression, the concrete and metal conduits are high stiffness, while the plastic conduit is low stiffness.

Other thermoplastic conduits are used for drainage conduit, including polyvinyl chloride (PVC) and polypropylene (PP). PVC has a higher short-term modulus than PE and the ratio of the long-term to short-term modulus is about 35% for PVC but 20% for PE. Because of this, PVC conduits will

Table 5.1 Conduits used in parameter studies with the Burns elastic model

Conduit Type	*Relative stiffness*	R_m *in. mm*	E_0 *ksi MPa*	E_{50}*ksi MPa*	PS_H *ksi MPa*	$PS_{B\text{-}US}$ *psi kPa*	$PS_{B\text{-}I}$ *psi kPa*
Concrete	High hoop, High bending	33 850	3,600 25,000		646,000 4,450,000	12,100 83,400	225 1,550
Metal (Corr. Steel)	High hoop, Low bending	30.25 770	29,000 200,000		62,000 430,000	13.3 92	0.17 1.1
Plastic (Corr. PE)	Low hoop, Low bending	31.5 800	110 760	22 150	290 2000	16 110	3.1 21

Note: In design, the long-term properties of polyethylene (PE) are used for hoop thrust calculations, and the short-term PE properties are used for deflection calculations. The basis for this is explained in Section 7.7.

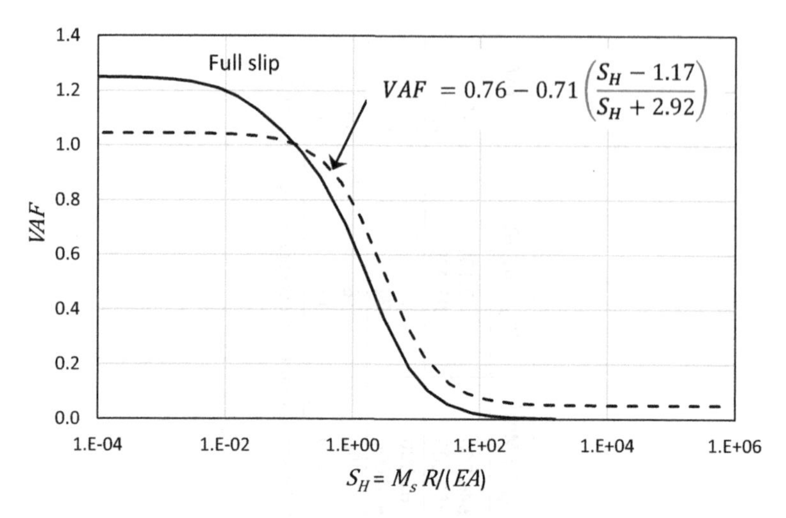

Figure 5.4 AASHTO VAF design curve for thermoplastic conduits (McGrath, 1999).

generally fall into the upper region of the semiflexible portion of the curve. PP conduits have properties similar to PE. Fiberglass conduits are also considered plastic but with a high modulus of elasticity and would plot in the same region as metal conduits in Figure 5.3.

McGrath (1999) noted the low *VAF*s possible in corrugated PE conduits and showed that the VAF could be computed based on the S_H ratio alone. He presented a simple expression to compute the VAF, which is consistent with the Burns curve for *VAF* and was adopted into the AASHTO LRFD Bridge Design Specifications (AASHTO 2017).

$$VAF = 0.76 - 0.71\left(\frac{S_H - 1.17}{S_H + 2.92}\right) \tag{5.5}$$

Figure 5.4 compares the Burns full-slip curve for *VAF* with Equation 5.5. The AASHTO VAF and the full-slip condition are not identical. The maximum portion was set at approximately 1.0 to match current practice, and the sloping portion was set slightly above the full-slip curve to provide some conservatism in design since the earth load on a thermoplastic conduit is sensitive to soil stiffness in the sloping portion of the curve.

5.1.2.2 Flexure

Conduit bending moment under earth load can be nondimensionalized by dividing it by $p_v R_m^2$, the soil prism pressure times the mean conduit radius squared. The parameter $M/p_v R_m^2$ plot versus S_B has the same characteristic shape as *VAF* versus S_H. Figure 5.5 shows the Burns estimates of crown and

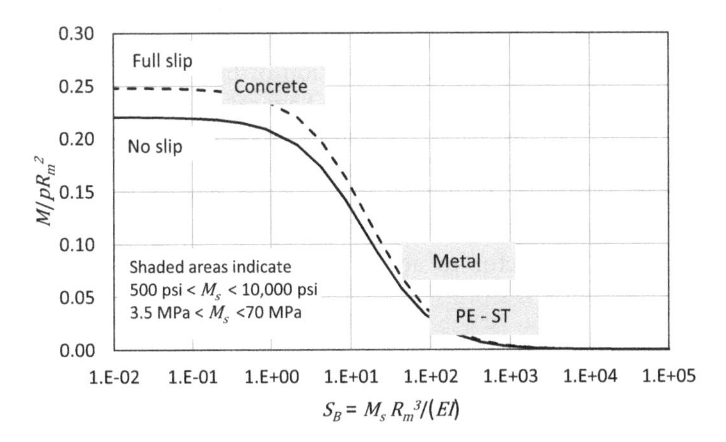

Figure 5.5 Bending moment versus bending stiffness ratio, S_B.

invert moments, which are identical because of symmetry. The shaded areas in Figure 5.5 show the range where the concrete, metal, and plastic conduits from Table 5.1 would plot for M_s varying from 500 to 10,000 psi (3,500 to 70,000 kPa).

The figure shows:

- The concrete conduit plots in the region where the response is controlled by the conduit stiffness; however, it suggests that some concrete conduits could be designed in the semirigid region of the curve. The analysis in Figure 5.5 is based on uncracked concrete conduit as is typical and conservative for design. Thin wall concrete conduits and many concrete conduits after cracking could be designed to take advantage of the soil support in this region, but the emphasis on installation quality would need to be increased.
- Typical metal conduits plot in the lower portion of the semiflexible region; thus, bending moments will be relatively small; however, shallow buried conduits and conduits with deep corrugations can be much stiffer; thus, some metal conduits are designed for bending moments, and some are not as discussed in Section 7.6.
- Plastic conduits plot in the same region as metal conduits. Thus, the figure demonstrates the traditional classifications of concrete conduits as rigid and metal and plastic conduits as flexible. Only the short-term condition for plastics is shown in Figure 5.5, as bending moments and deflection mostly occur during the construction stage and shortly thereafter.

Figure 5.5 demonstrates the general behavior of conduits but is not useful for design because it fails to address bending moments due to nonuniform support (e.g., bedding and haunch zones) and nonuniform load conditions (e.g., live loads). Metal conduits with corrugations two inches deep or less are typically designed without consideration of bending moments. This is in part because these culverts can tolerate flexural stresses beyond the yield stress and in part because deflection and shape distortion limits imposed during construction are specified to keep stresses at acceptable levels. Plastic conduit design often involves calculation of the deflection level and then computing the bending moments based on the parallel plate test equations modified by a shape factor to account for nonuniform support. This is addressed in Section 7.7

5.1.3 Soil stresses around a conduit

The Burns and Richard solution also provided equations to determine the interface stresses on a conduit and in the soil mass away from the conduit. Figure 5.6 presents the calculated interface pressures on the conduits from Table 5.1. The soil pressures are divided by the soil prism pressure to present nondimensional scales. The graphs presented are for the typical interface design assumptions – that is, no-slip for concrete and full-slip for the flexible conduits: metal and plastic. For the plastic conduit, the interface pressures are shown for both the short- and long-term conditions to demonstrate changes that can occur over time. Two soil stiffnesses are presented to represent a range of soil types.

Figure 5.6 shows the following:

- Concrete conduit – with both high bending and high hoop stiffness the conduit undergoes minimal deflection and thus does not develop significant lateral pressures beyond those due to the naturally occurring K_o condition, as demonstrated by the significant difference between pressure at the crown and invert relative to the pressure at the springlines. The radial pressures on the conduit are relatively insensitive to the soil stiffness.
- Metal conduit – With low bending stiffness but high hoop stiffness, the metal conduit deflects, allowing the transfer of load off the conduit, resulting in lower crown pressures relative to the concrete conduit. As a result of the deflection, lateral pressures beyond the K_o condition develop, resulting in a relatively uniform pressure distribution around the conduit. Like the concrete conduit, radial pressure on the metal conduit is relatively insensitive to changes in soil stiffness.
- Plastic conduit – With low bending and low hoop stiffness, the plastic conduit shows notable changes relative to the concrete and metal conduits. Like the metal conduit, plastic has a relatively uniform pressure distribution around the circumference. Unlike the metal and concrete conduits, the radial pressures on the plastic conduit show a significant sensitivity to the soil stiffness. High soil stiffness reduces the load applied to the conduit as the conduit shortens circumferentially due to

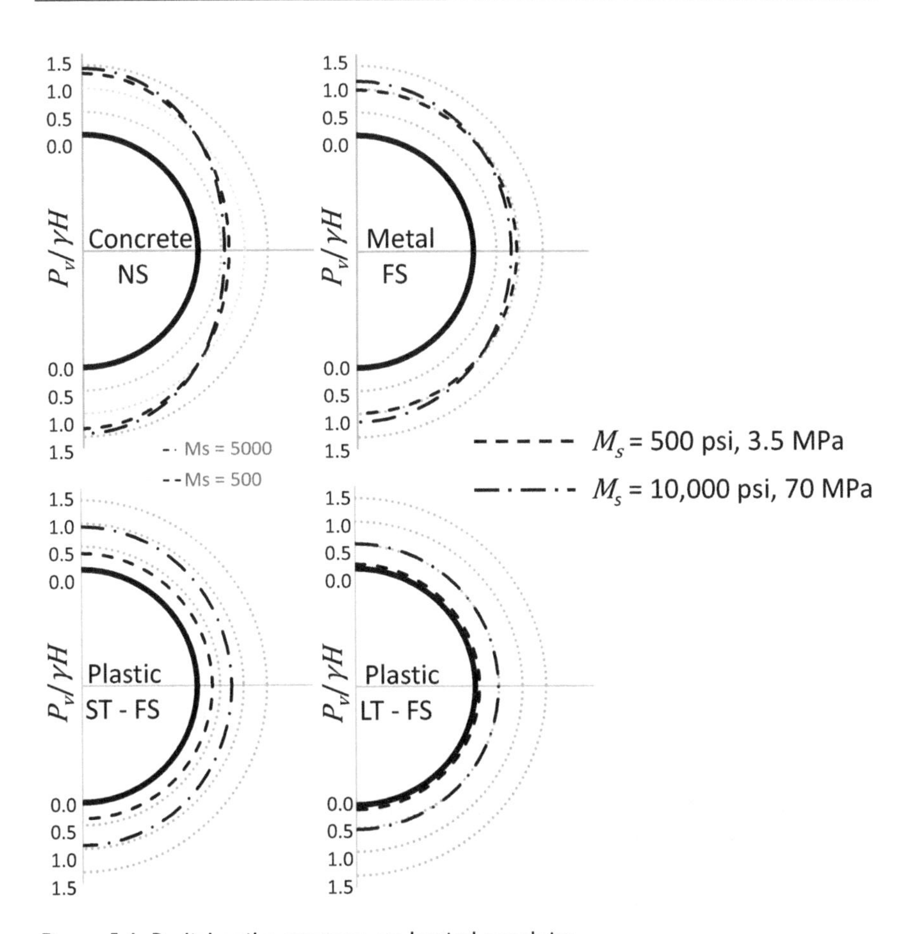

Figure 5.6 Radial soil pressures on buried conduits.

the low hoop stiffness. Further, when the calculations are made using the long-term modulus of elasticity, the radial pressures drop even further. This latter effect represents changes that occur over time after installation.

Figure 5.7 presents the tangential shear stresses at the interface of a concrete conduit for the no-slip condition. Flexible conduits are designed for the full-slip condition where there are no interface shear stresses. In the figure, negative shear stresses are in the clockwise direction (downward) on the right side of the conduit, and positive shear stresses are counterclockwise (upward) on the right side of the conduit. It is the presence of interface shear stresses that create the significant increase in hoop thrust at the springline.

Another finding of the Burns solution is that the soil stresses around a buried conduit attenuate to near the free field soil condition by one diameter

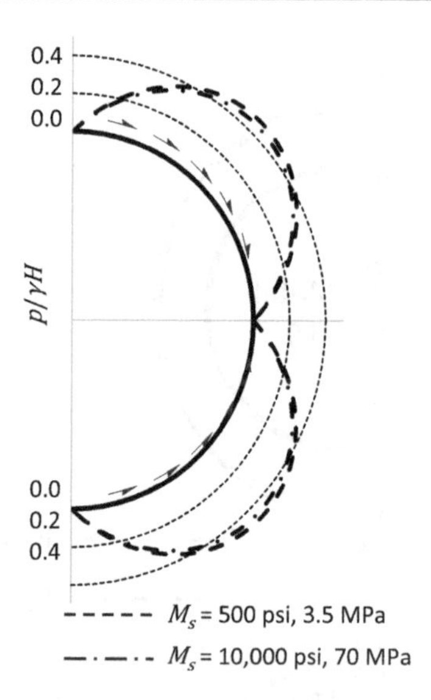

Figure 5.7 Soil shear stresses on concrete conduit interface.

from the soil-conduit interface – behavior that has significant implications for whether conduits are affected by other structures (e.g., other conduits).

5.2 EARTH PRESSURE DISTRIBUTION AND LOAD THEORIES FOR DESIGN

The benefits of high lateral soil pressure and uniform bedding support are key in limiting thrusts, moments, and displacements, and thus in reducing the cost of a conduit. For flexible conduits, the only widely used simplified distribution is the Iowa formula developed by Spangler and introduced in Chapter 2. For rigid conduits, three distributions have been used in addition to that proposed by Marston, which was also introduced in Chapter 2. The current chapter discusses all these approaches and their effectiveness in addressing the important design issues of soil-conduit interaction.

5.2.1 Flexible conduits – Spangler pressure distribution

The Spangler pressure distribution (Figure 5.8) was published by Iowa State (Spangler, 1941) after many years of field and laboratory tests conducted on flexible corrugated metal conduits.

The benefit of lateral pressure to reduce forces in rigid conduit was known, but rigid conduit design used at rest pressure, while flexible conduit design required a method to address the extra pressure developed when a

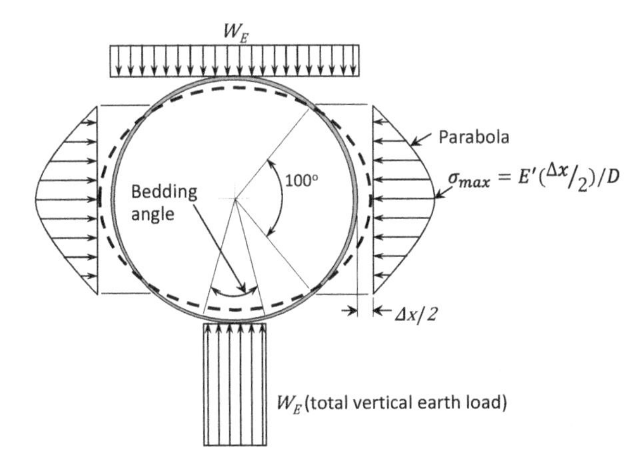

Figure 5.8 Spangler pressure distribution.

conduit deflects into the soil. The Spangler distribution addressed this by making the lateral pressure a function of the conduit deflection into the soil. This allowed the development of the soil-conduit interaction equation, now called the Iowa formula or the Spangler equation:

$$\frac{\Delta_y}{D_m} = \frac{K_x\left(D_l\, W_E + W_L\right)}{\dfrac{EI}{R_m^3} + 0.061\, E'}. \tag{5.6}$$

The soil stiffness, E', or M_s, is the key parameter in controlling deflection. Soil stiffness was discussed in Section 4.3. Several other terms in Equation 5.6 require discussion, which follows the definition of the newly introduced terms:

Δ_y/D_m = Change in horizontal diameter as a fraction of undeformed diameter
K_x = Bedding factor
D_l = Deflection lag factor
W_E, W_L = Earth and live load, lb/in., kN/m (SI units are selected to keep the equation consistent)
EI/R_m^3 = conduit stiffness, psi, kPa
E' = modulus of soil reaction, psi, kPa

- Δy – As noted, the Iowa formula was developed to predict change in the horizontal diameter of a circular conduit. However, much of the data collected to evaluate installations has been the change in vertical diameter, as it is simpler to measure. As a result, the formula is often, if not mostly, used to predict change in vertical diameter.

- D_l – The deflection lag factor accounts for change in diameter after the installation is completed. This time-dependent change occurs as the soil around the conduit settles and densifies, largely due to water infiltration and cyclic vehicle loads. Spangler, based on limited data, recommended a "conservative" design value of 1.5. Howard (2015) presents values that range from 1.5 to 3.0. The higher values in Howard's table are all associated with stiff backfills, which will have small calculated deflections; thus, the larger values of the deflection lag factor may not be significant. For this reason, some standards, such as the AASHTO *LRFD Bridge Design Specifications* (called AASHTO LRFD, AASHTO, 2020) and AWWA *Manual M45 for Fiberglass Pipe* (AWWA, 2014) recommend a single value of 1.5.
- W_L – Live loads are temporary loads. For conduits, this mostly refers to vehicle loadings discussed in Section 5.3 and Chapter 7.
- K_x – The bedding factor adjusts the horizontal deflection due to changes in the width of the bedding support at the bottom of the conduit. However, there are several shortcomings in using this factor:
  - The horizontal pressure distribution is not adjusted for narrow bedding, yet it is likely that a poorly installed conduit with narrow bedding will have reduced lateral support in the lower half of the conduit, as well as a narrower bedding width.
  - Spangler's bedding factor varies from 0.083 for full-width, 180° bedding to 0.110 for 0° bedding (a line load), a variation of 30%. This represents the result of poor bedding on the horizontal deflection. It is likely that poor bedding will have a greater effect on the vertical deflection.

Overall, the Spangler pressure distribution and Iowa formula are useful for evaluating the relative effects of varying backfill types and densities during the design phase. This calculation should be considered a guide showing the general feasibility of a proposed design. The best control of deflection is achieved through the use of appropriate backfill materials and approved construction methods. Monitoring deflection after installation is an important step in achieving this, as discussed in Chapters 8 and 9.

5.2.2 Rigid conduits

Several pressure distributions have been devised to assist in the analysis of buried rigid conduits. Often referred to by their developer's names, they include the uniform load system (Paris, 1921), the radial load system (Olander, 1950), and later, the standard installation distribution (Heger, 1988). Analysis with these distributions results in the forces that can be used to design reinforcement directly for in-ground conditions. This is called direct design in AASHTO LRFD and is discussed in Section 7.5.2.

5.2.2.1 Radial (Olander 1950) and Uniform (Paris 1921) pressure distributions

The uniform and radial pressure distributions are shown schematically in Figure 5.9 for the conditions of conduit weight, soil weight, fluid load, and live load. Both methods provide coefficients that are used to calculate internal forces resulting from the applied loads.

The radial load system (Figure 5.9a) applies soil pressure perpendicular to the conduit surface, varying in sinusoidal fashion around the circumference. As originally developed by Olander, the sum of the earth and bedding pressure angles was 360°, which means decreasing bedding angles increase the lateral earth pressure on the conduit. This is not logical for small bedding angles, as narrow bedding typically results in a void in the haunch zone and so should reduce lateral pressure below the springline. Thus, the radial pressure distribution was not appropriate for bedding angles of less than 90°. McGrath, Tigue, and Heger (1988) developed the computer program PIPECAR for the design of concrete conduit and allowed the sum of the load and bedding angles to be less than 360° (Figure 5.9, 2a). This allowed an unloaded area

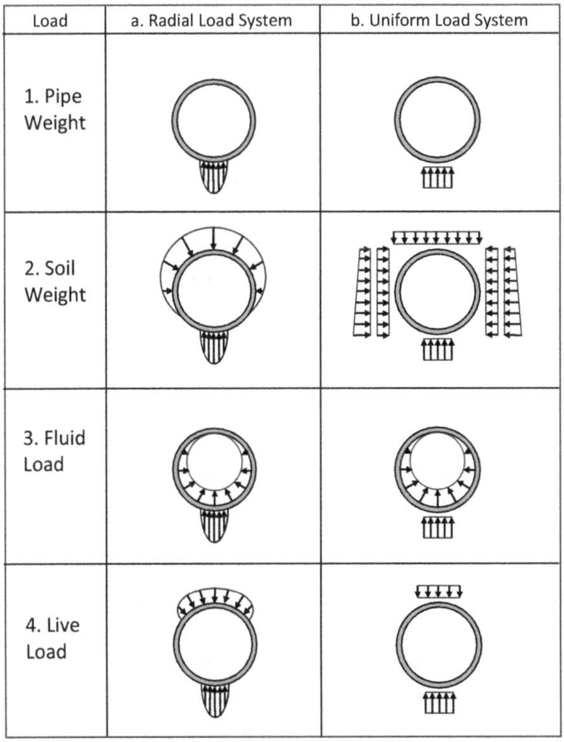

Load	a. Radial Load System	b. Uniform Load System
1. Pipe Weight		
2. Soil Weight		
3. Fluid Load		
4. Live Load		

Figure 5.9 Radial and uniform pressure distributions. (ACPA 1998, adapted with permission)

between the bedding and load angles, which is consistent with field testing. As of this writing, PIPECAR has been replaced with Eriksson Pipe (Ericksson, 2021), an updated version of PIPECAR, which is supported by the American Concrete Pipe Association, the concrete conduit trade organization. Concrete pressure conduit designs in ASTM C361 *Standard Specification for Reinforced Concrete Low-Head Pressure Pipe* are analyzed using the radial pressure distribution with 90° bedding and 270° load angle.

The uniform load system was initially developed by Paris as design coefficients for various individual loads, which could then be combined to create a desired loading condition. For example, the soil loading condition (Figure 5.9, 2b) would be made up of three conditions – the vertically applied pressures, uniform lateral pressure, and a tapered lateral pressure. In its original form, a table of moment coefficients was developed for each individual condition. Now the separate conditions may be combined directly in a computer program.

The radial and uniform distributions were widely used prior to the advent of computer analysis. They are still used in some design standards for concrete conduits. Both methods are available in Erikson Pipe.

5.2.2.2 Heger distribution

Using the finite element computer program SPIDA (Heger et al., 1985), Heger (1988) proposed a pressure distribution more complex than the Paris and Olander approaches and one that addressed a wider range of installation conditions. This pressure distribution formed the basis of the four "standard installations" incorporated into AASHTO LRFD for concrete conduits. These pressure distributions were developed for use in computer programs without the necessity of implementing finite element analysis. The four installations vary from Type 1, in which the sidefill below the springline is a coarse-grained soil compacted to 95% of maximum standard Proctor density, including compacted backfill in the haunch zone, to Type 4, which has minimal backfill and compaction requirements.

The AASHTO LRFD Bridge Construction Specifications (AASHTO 2017, henceforth called AASHTO Construction) and industry literature (ACPA 1998) identify the important backfill zones around a concrete conduit (Figure 5.10):

- The middle bedding area under the central third of the conduit diameter is loosely placed for Type 1, 2, and 3 installations. This allows a slight settlement during backfilling, which increases the amount of earth load carried by the haunch backfill relative to the bedding. Type 4 installations, which have minimal requirements for backfill type or compaction level, do not require loose middle bedding.
- The haunch zone in Figure 5.10 encompasses the structural backfill at the side of the conduit as well as the triangular area below the

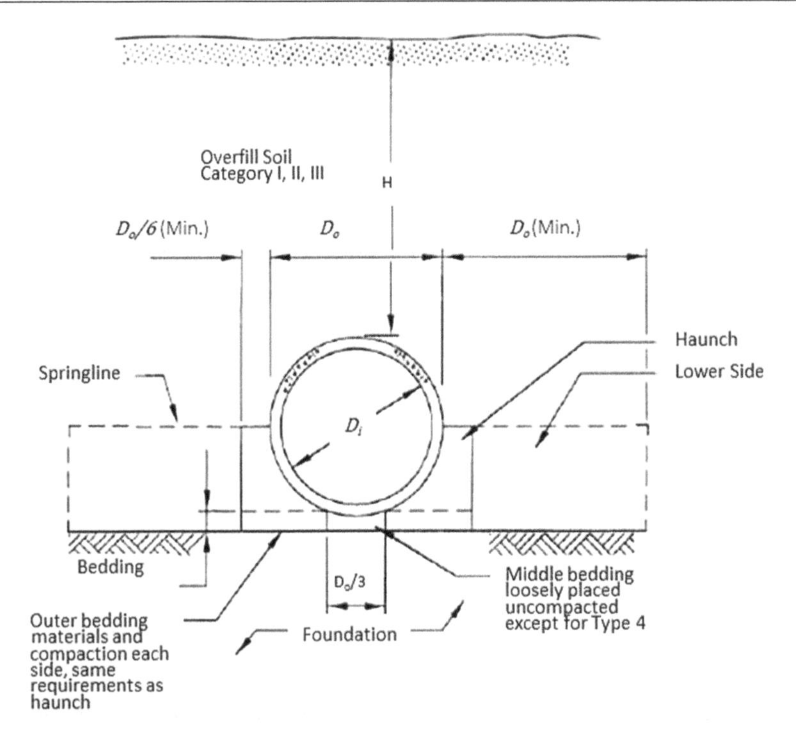

Figure 5.10 Standard embankment installation. (ACPA 1998, adapted with permission)

springline, which is different from common terminology for flexible conduits defined in Figure 2.2a. The backfill in the haunch zone must be worked into the area under the springline and firmly compacted, ideally to the same density as required for the sidefill, although this is not always fully achievable. Compaction in the haunch zone is required for Type 1, 2, and 3 installations. Type 4 installations have minimal backfill requirements in this zone.

- The lower side zone is part of the embankment fill and not part of the structural backfill, but it must have sufficient strength and stiffness to carry vertical earth load. If the soil in this zone is too compressible the VAF will increase above the design value.
- Above the springline, there are minimal backfill and compaction requirements for conduit support, but control of backfill material and density may be necessary if the conduit is installed under a roadway.

The Heger distribution (Figure 5.11) includes areas of pressure, A1 to A6, that vary in size as a function of the installation type. Coefficients defining the load represented by each area are provided in Table 5.2. The vertical arching factor, *VAF* (called F_e in AASHTO), and horizontal arching factor, *HAF*, represent the nondimensional total vertical and horizontal load on

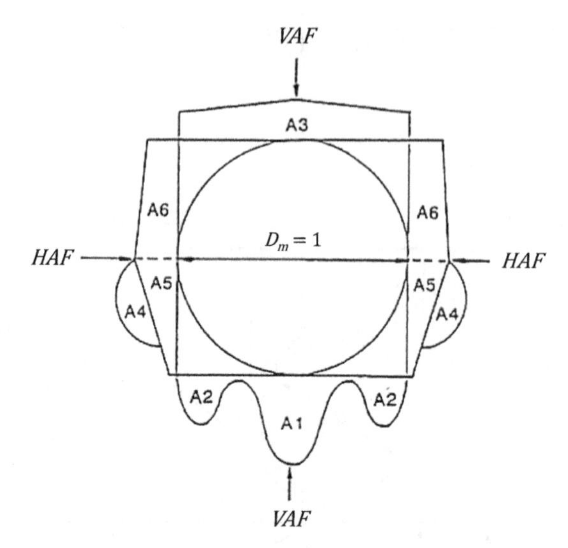

Figure 5.11 Heger pressure distribution. (Heger 1988, with permission from ASCE)

Table 5.2 Load coefficients for standard installations (Heger, 1988 with permission from ASCE)

Installation Type	*VAF*	*HAF*	*A1*	*A2*	*A3*	*A4*	*A5*	*A6*
1	1.35	0.45	0.62	0.73	1.35	0.19	0.08	0.18
2	1.40	0.40	0.85	0.55	1.40	0.15	0.08	0.17
3	1.40	0.37	1.05	0.35	1.40	0.10	0.10	0.17
4	1.45	0.30	1.45	0.00	1.45	0.00	0.11	0.19

Note: For clarity, only the total load within each pressure zone is shown here. Additional coefficients in AASHTO LRFD fully define the pressure distribution.

the conduit relative to the soil prism load, i.e., $W_E = VAF\ (\gamma_s\ H\ D_o)$ and total horizontal load, W_H, equals $HAF\ (\gamma_s\ H\ D_o)$. The downward load on the conduit, A3 is in equilibrium with the upward load, A1 + A2. Similarly, the horizontal load, A4 + A5 + A6, is identical on both sides of the conduit. AASHTO LRFD and ACPA (1993) provide additional information to define the exact shape of the pressure distribution. Analysis and design of concrete conduits using the Heger distribution is incorporated into Erikson Pipe (2021).

The Heger pressure distribution is most easily understood by plotting the pressure distribution for the four standard installations to scale (Figure 5.12), which demonstrates the following:

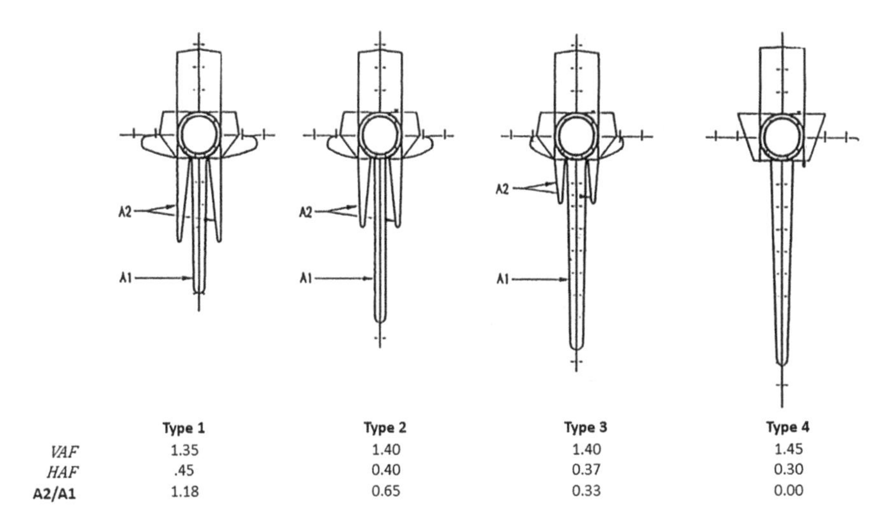

	Type 1	Type 2	Type 3	Type 4
VAF	1.35	1.40	1.40	1.45
HAF	.45	0.40	0.37	0.30
A2/A1	1.18	0.65	0.33	0.00

Figure 5.12 Heger pressure distribution drawn to scale.

- Installation Types 1, 2, and 3, which feature reduced compaction in the middle bedding, have lower pressures at the bottom of the conduit.
- Type 1 installations have the highest lateral load (HAF), and over 50% of the vertical load is carried by backfill in Zone A2.
- As the sidefill and compaction requirements are relaxed from Type 1 to Type 4 installations, the vertical load in Zone A1 increases relative to A2, and the lateral load below the springline drops. In a Type 4 installation, both Zones 2 and 4 are zero.

Chapter 7 on design shows the significant increase in bending moments in the conduit as the installation type changes from Type 1 to Type 4.

5.3 VEHICLE LOADS ON BURIED CONDUITS

Many conduits are installed under roadways, parking lots, railroad lines, and airport taxiways and runways. Vehicle loads can be the dominant portion of the total load on shallow buried conduits but attenuate rapidly as the depth below the surface increases and at some depth become insignificant.

Boussinesq developed one of the classic equations for distributing a surface point load through a uniform solid (Equation 5.7). The load is applied as a point load, and the spread through the solid is axisymmetric. Figure 5.13 shows the vertical soil stress profile, σ_z, at several depths of fill calculated in accordance with Equation 5.7. The Boussinesq equation is

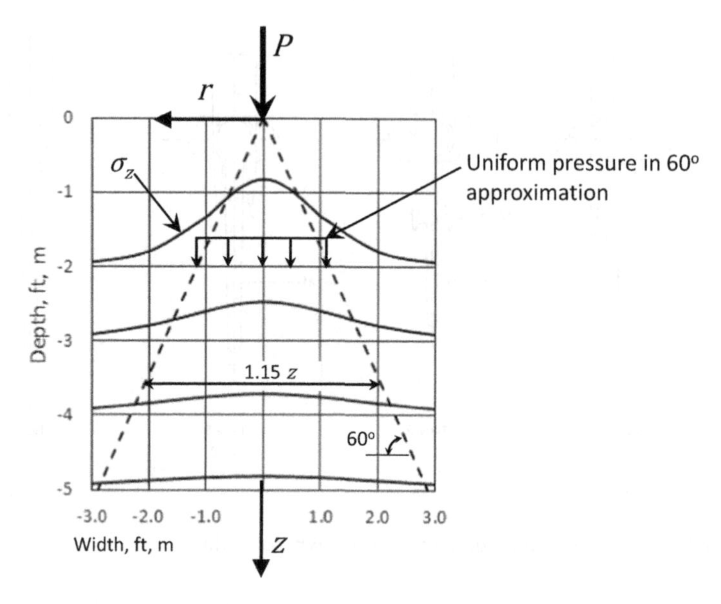

Figure 5.13 Boussinesq load distribution and 60° approximation.

widely known and useful for introducing the subject of live load distribution through fills, but design methods for live load distribution require a distributed load at the surface representing the tire contact area, as introduced later.

$$\sigma_z = \frac{3P}{2\pi z^2 \left[1 + \left(\frac{r}{z}\right)^2\right]^{5/2}} \tag{5.7}$$

Figure 5.13 also shows a common simplification used for distributing concentrated loads through fill. In this approximation, the entire concentrated load is distributed as a uniform pressure over a width equal to 1.15 times the depth of fill. The rate of distribution is called the live load distribution factor, LLDF. The angle of the line showing this spread is approximately 60° relative to the horizontal axis. A uniform distribution is simpler for analysis, and even though the average pressure is less than the peak pressure, the structural response of the conduit is sufficiently accurate for design. Distributing live loads in this manner is the primary method in AASHTO LRFD, as demonstrated in Figure 5.14 and many other design codes. AASHTO LRFD includes some additional modifications to live load distribution based on the hoop and bending stiffness ratios discussed earlier. These are addressed in Chapter 7.

Figure 5.14 is drawn with a tire footprint of l_t = 0.83 ft (0.25 m) in the travel direction and w_t = 1.67 ft (0.5 m) in the transverse direction, which represents the typical dual tire configuration for a truck. Typical center-to-center dual wheel spacing on one axle is 6 ft (1.8 m).

When the loads spreading from adjacent wheels of an axle intersect, the entire load of both wheels is distributed as a uniform pressure over the total area of the intersecting loads, as shown in Figure 5.15 (shown only in US units for clarity). The width of the distribution of the wheel loads is, as shown earlier, to a depth of 3.8 ft (1.15 m), where the distribution of the two wheel loads interact. At greater depths, the effect of the total load, $2W_t$, is distributed over the total width of the two loads, 6 + 1.67 ft + 1.15 H (1.8 m + 0.5 m + 1.15 H). The same distribution method is used in the traffic direction. Wheel loads on tandem axles, which are typically spaced at 4 ft (1.2 m), interact at a depth of 2.75 ft when two wheels on adjacent axles interact and later when wheels on the same axle interact. AASHTO LRFD requires checking a 4 ft (1.2 m) center-to-center spacing of adjacent wheels when evaluating two trucks in adjacent lanes; however, as discussed next, this is not typically a controlling load condition for conduits.

While truck configurations and axle loads can vary widely, railroads take a different approach. The American Railway Engineering and Maintenance-of-Way Association (AREMA), which issues the Manual for Railway Engineering (2019), designs buried conduits for a locomotive load, typically the Cooper E80 load which, for the purpose of designing concrete box conduits, is a continuous series of 80 k (350 kN) axles, spaced at 5 ft (1.5 m). The locomotive load is distributed transverse to the tracks as a uniform load through ties, which are most commonly 8 ft 6 in. (2.6 m)

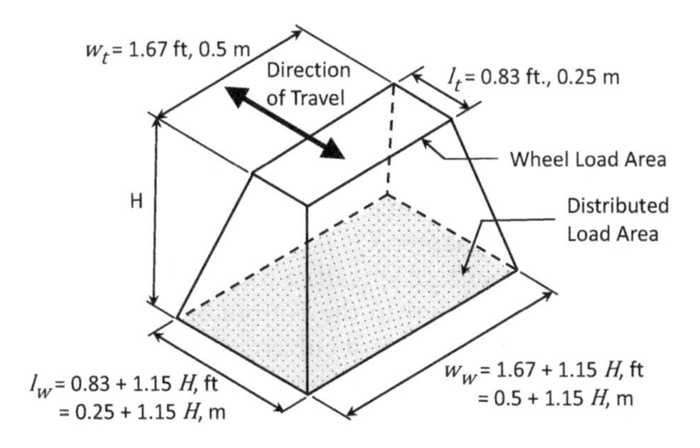

Figure 5.14 AASHTO live load distribution through soil for a typical highway truck wheel.

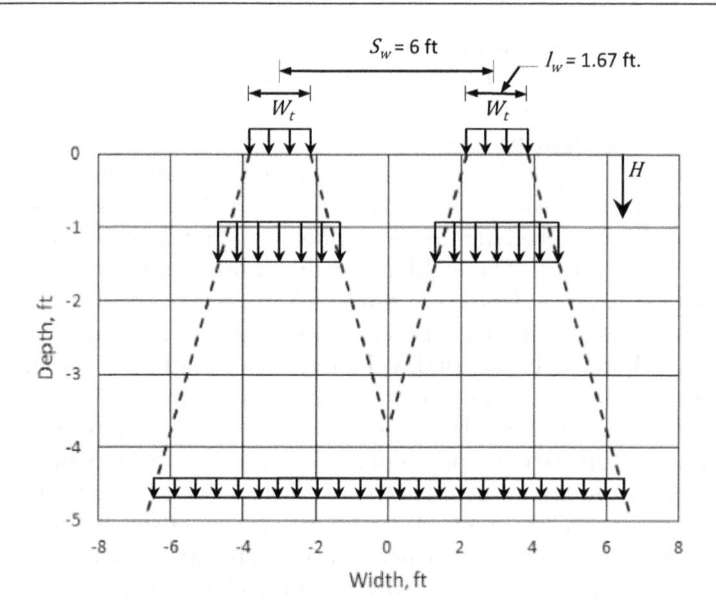

Figure 5.15 Distribution of intersecting wheel loads through fills.

long, with LLDF = 1.0. The Manual also considers track weight to be 0.2 k/ft (2.9 kN/m) and is also distributed only in the transverse direction.

Live loads from tracked vehicles, such as excavators, can be distributed through fill in the same manner as trucks. Tracked vehicles typically have low contact pressure relative to trucks but a much larger loaded area. Thus, the attenuation of load with depth will be slower and the live load may be significant at greater depths than trucks.

Aircraft live loads are typically distributed in a fashion similar to truck loads. The wheel configurations and loads are widely variable. Wheel loads of 50,000 lb (220 kN) are common design loads for smaller craft, but wide-body jets have total loads up to about 1,000 k (4,400 kN) distributed over several wheel groups.

AASHTO LRFD presents detailed equations for live load calculations in Article 3.5.1.2.6. Additional details for calculating the live load applied to conduits are presented in Chapter 7.

A pavement on a road surface spreads a live load over a larger area than the wheel footprint. The pressure at the bottom of the pavement is reduced but is applied over a larger area. Most culverts are designed without consideration of the improved load distribution resulting from pavements over a conduit. The only exception to this is some metal box section designs as detailed in AASHTO LRFD Article 12.9.4.6. The effect of pavements is ignored primarily to allow for construction loads prior to placement of pavement and perhaps to allow for a hot asphalt pavement where the modulus can drop significantly. Finite element analysis can be used to model

Table 5.3 Pavement effect in distributing live load on culverts

Pavement Thickness, in.	*Asphalt Stiff Subgrade*	*Asphalt Soft Subgrade Concrete Stiff Subgrade*	*Concrete Soft Subgrade*
	E1/E2 ~3	*E1/E2 ~35*	*E1/E2 ~400*
4	NB	NB	0.50/5 ft (1.5 m)
8	NB	0.60/6 ft 1.8 m)	0.25/6 ft (1.8 m)
16	0.75/6 ft (1.8 m)	0.50/7 ft (2.1 m)	0.15/8 ft (2.4 m)

Where
*E*1 = modulus of pavement layer
*E*2 = modulus of soil subgrade
NB = no benefit

The data lines, such as 0.50/5 ft (1.8 m), indicate a reduction factor that may be applied to the live load at the bottom of the pavement and the depth at which no benefit is derived in reducing pavement load.

Source: Mlynarski et al., 2019. Copyright National Academy of Sciences. Reproduced with permission of the Transportation Research Board.

pavement effects; however, most finite element analysis of conduits is completed using two-dimensional models, which will not capture the full effect of live load spread in three dimensions. Empirical procedures for considering pavements include elasticity theory procedures for layered systems and the Westergaard procedure for distributing live loads through concrete pavements, as embodied in the *Concrete Pipe Handbook* (ACPA, 1998).

Considering the load reduction from pavement can be important in load rating culverts where it can make the difference between finding a culvert capacity to be adequate or requiring strengthening or replacement. Table 5.3 presents guidance on the conditions and locations where pavements are effective in reducing loads on culverts.

Table 5.3 is used by conducting a typical live load calculation with the standard wheel footprint at the top of the pavement and then reducing the pressure at the bottom of the pavement using the table reduction factor, increasing the loaded area to maintain the same total load. Use the LLDF to spread the load at depths below the pavement. The reduction factor should be varied linearly to zero with increasing depth.

Table 5.3 was derived from an elastic solution by Fox and presented in Poulos and Davis (1991) using the following assumptions:

- *E*-concrete pavement = 4,000 ksi (28,000 MPa)
- *E*-asphalt pavement = 300 ksi (8.4 MPa)
- *E*-soft subgrade approximately 8 ksi (56 MPa)
- *E*-stiff subgrade approximately 100 ksi (700 MPa)

Two relationships between the soil modulus and the common parameters, as recommended by the Federal Aviation Administration Advisory Circular 150/5320-6F, 2016, are

$$E = 1{,}500\,\text{CBR}, \tag{5.8}$$

$$E = 20.15\,k^{1.284}, \tag{5.9}$$

where

E = modulus of elasticity of subgrade, psi
CBR = California bearing ratio
k = modulus of subgrade reaction, pci

Note that Equations 5.8 and 5.9 provide modulus values considerably higher than typically used in culvert backfill design.

As an example, for an 8-in. concrete pavement with a soft subgrade, the live load could be reduced to 25% of the applied load for a culvert directly under the pavement, and there would be no reduction if the culvert is more than 5 ft below the pavement. Linear extrapolation can be used to determine the reduction for intermediate depths.

5.4 SPECIAL ASPECTS OF SOIL-CONDUIT INTERACTION

5.4.1 Load reduction techniques

Various procedures have been developed to reduce earth loads on buried conduits. These include negative projecting and induced trench installations introduced in Chapter 2, soft bedding, and lightweight fills. Even installing conduits in narrow trenches provides a form of load reduction. Reducing load on a conduit requires redirecting the load path, as shown in Figure 5.16. See Section 9.1 for a case where load path was not considered.

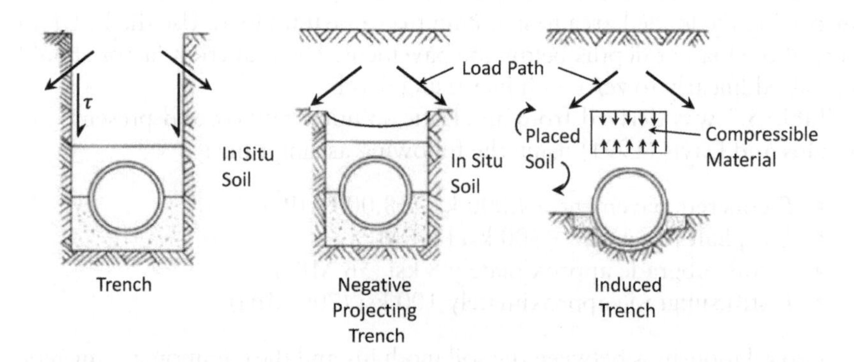

Figure 5.16 Load path to reduce earth load on buried conduit.

Each of the installations in Figure 5.16 uses a different load path to transfer earth load from the conduit to the adjacent soil:

- The trench load depends on settlement/consolidation of the trench backfill after installation to create forces on the trench wall that transfer part of the backfill into the in situ soil. The narrower the trench, the more effective the shear stresses are in transferring trench backfill load to the trench wall. If the trench width is too large, the wide trench condition discussed in Chapter 2 (see Figure 2.7) will be in control, and there will be no load reduction relative to the embankment condition. The trench wall must have sufficient strength and stiffness to support the additional load. If the trench wall stands without bracing or a trench box, this will normally be the case; however, a trench box or other trench support is almost always required for worker safety. Moving a trench box after placing sidefill can be disruptive to the compaction and can affect the trench load calculation. See Chapter 8 for additional discussion of trench boxes.

  While narrow trenches are beneficial for load reduction, designers must remember that the trench still needs to be wide enough to allow placing and compacting backfill in the haunch zone and at the sides of the conduit.
- The negative projecting trench condition depends on the settlement/consolidation of the subtrench backfill to transfer the weight of soil over the conduit into the in situ soil at the sides of the subtrench. The mechanisms for this transfer are shear forces that develop within the embankment material over the conduit due to backfill settlement in the subtrench. The load transfer becomes more effective as the depth of the subtrench increases.
- In the induced trench installation, the conduit is placed on native material, and the embankment is built up around and over the conduit. After reaching an appropriate height over the conduit, a subtrench is excavated over the conduit and filled with compressible material that causes the transfer of load to the embankment material at the sides of the subtrench. The embankment material at the sides of the conduit must be sufficiently stiff to carry the additional load with minimal compression. Although it appears obvious, the subtrench must be located directly over the conduit, which may not be visible during excavation. The induced trench installation is not widely used.

In all of these cases, the soil receiving the load transferred off the conduit must be sufficiently stiff and strong to accept this load transfer without excessive compression, which would reduce the amount of load that can be transferred. Most native soils that are stable during trench excavation meet this criterion, but each project must be evaluated by the design engineer.

An additional method of load reduction is the use of lightweight fills over the conduit. Materials such as expanded polystyrene and expanded clay shale have been used for this purpose.

5.4.2 Compaction effects during construction

The compactibility of soils – that is, the energy required to densify soils, was discussed in Section 4.3.4. Another aspect of compaction effects is the interaction with the conduit next to the soil being compacted. Compaction equipment achieves densification by applying the weight of the compactor, commonly augmented by a dynamic force created by vibration or impact. Next to flexible conduits, these forces can lead to deformations that, in the extreme, can be detrimental to long-term performance. Installation specifications for flexible conduits sometimes impose a limit on upward deflection during placement and compaction of embedment soil to prevent excessive compaction deformation. This limit is less than the downward deflection limit, as compaction effects can deform conduits into nonelliptical shapes. Long-span structural plate structures, as defined by AASHTO LRFD, are particularly flexible and subject to this type of deformation. Steps are taken in design (see Section 7.6.2) and during construction to control this deformation. See Section 9.3 for an extreme example of monitoring for compaction effects but not responding properly to the data provided by the inspection.

There is no widely accepted model for analyzing soil-conduit interaction resulting from compaction. Capturing the complex soil behavior, which includes repeated stress cycles, reduction of voids, and plastic soil deformations, is challenging and not achievable with the soil models in most common use for conduit design. McGrath et al. (1998) evaluated the compaction effects of a vibratory plate and an impact compactor (jumping jack) used to densify a stone backfill and a silty sand backfill (Soils 3 and 6 in Figure 4.2) around 36 and 60 in. (900 and 1,500 mm) diameter corrugated polyethylene and corrugated steel conduits. In addition to the obvious difference in applied vertical stress due to compaction, McGrath found that the horizontal stresses applied to a conduit during compaction are a function of $(1 - \sin \phi)^3$, which means that compacting the silty sand ($\phi \sim 28°$) with the same compactor applied approximately twice the horizontal stress on the conduit relative to the stone backfill ($\phi \sim 36°$).

5.4.3 Longitudinal loading

Buried conduits carrying gravity flow fluids need to be installed with a uniform grade to provide proper drainage. With this goal and with the need to limit stresses in the conduits, specifications call for uniform longitudinal bedding support to minimize and hopefully eliminate longitudinal bending of conduit sections. If such support is not provided, conduits will act as beams to span the unsupported lengths and "beam breaks" may result.

Smaller diameter conduits are most susceptible to this as they have high ratios of the length of a single piece of conduit to its diameter, and so peak longitudinal bending moments are much higher fractions of longitudinal bending strength. Assessment of the in situ soils and proper placement of bedding and foundation soils are important steps in providing uniform support. Dense compaction of backfill in the haunch zone is also important in avoiding beam breaks as the vertical haunch support can provide support that will prevent the conduit from settling over uneven bedding.

REFERENCES

AASHTO (2017) *AASHTO LRFD Bridge Construction Specifications*, 4th Edition, AASHTO, Washington, DC.

AASHTO (2020) *AASHTO LRFD Bridge Design Specifications*, 9th Edition, AASHTO. Washington, DC.

ACPA (1993) Concrete *Pipe Technology Handbook*, American Concrete Pipe Association, Irving, TX.

ACPA (1998) *Concrete Pipe Handbook*, American Concrete Pipe Association, Irving, TX.

AREMA (2019) *Manual for Railway Engineering*, American Railway Engineering and Maintenance-of-Way Association, Lanham, MD.

AWWA (2014) *Fiberglass Pipe Design, Manual of Water Supply Practices—M45*, 3rd Edition, American Water Works Association, Denver, CO.

Burns, J.Q., and Richard, R.M. (1964) Attenuation of Stresses for Buried Circular Cylinders, *Proceedings of the Symposium of Soil Structure Interaction*, University of Arizona, Tucson, AZ, pp. 378–392.

CANDE (2022) *Culvert Analysis and Design*, https://www.candeforculverts.com/home.html

Ericksson (2021) *Eriksson Pipe*, Eriksson Software Inc., Temple Terrace, FL.

Heger, F.J. (1988) New Installation Designs for Buried Concrete *Pipeline Infrastructure – Proceedings of the Conference*, American Society of Civil Engineers, New York, NY, pp 117–135.

Heger, F.J., Liepins, A.A., and Selig, E.T. (1985) SPIDA Analysis and Design System for Buried Concrete Pipe, *Proceedings: Advances in Underground Pipeline Engineering*, American Society of Civil Engineers, Reston, VA, pp. 143–154.

Howard, A.K. (2015) *Pipeline Installation 2.0*, Relativity Publishing, Lakewood, CO.

Katona, M.G., Smith, J.M., Odello, R.S., and Allgood, J.R. (1976) *A Modern Approach for the Structural Design and Analysis of Buried Culverts, FHWA Report No. FHWA-RD-77-5*, Federal Highway Administration, Washington, DC. candeforculverts.com

McGrath, T.J., Tigue, D.B., and Heger, F.J. (1988) *PIPECAR and BOXCAR – Microcomputer Programs for the Design of Reinforced Concrete Pipe and Box Sections (Transportation Research Record No. 1191)*, Transportation Research Board, Washington, DC.

McGrath, T.J., Selig, E.T., Webb, M.C., and Zoladz, G.V. (1998) *Pipe Interaction with the Backfill Envelope, FHWA-RD-98-191*, Federal Highway Administration, Washington, DC.

McGrath, T.J. (1999) *Calculating Loads on Buried Culverts Based on Pipe Hoop Stiffness (Transportation Research Record, No. 1656)*, The Transportation Research Board, Washington, DC.

Mlynarski, M, McGrath, T.J., Clancy, C., and Katona, M.G. (2019) *Proposed Modifications to AASHTO Culvert Load Rating Specifications (NCHRP Web-Only Document 268)*, National Cooperative Highway Research Program. Washington, DC.

Olander, H.C. (1950) *Stress Analysis of Concrete Pipe, Engineering Monograph No.* 6, Bureau of Reclamation, US Department of the Interior. Denver, CO.

Paris, J. M. (1921) Stress Coefficients for Large Horizontal Pipes, *Engineering News Record*, Vol. 37, No. 19.

Poulos, H.G., and Davis, E.H. (1991) *Elastic Solutions for Soil and Rock Mechanics*, http://usucger.org

Spangler, M.G. (1941) *The Structural Design of Flexible Pipe Culverts (Iowa Engineering Experiment Station, Bulletin 153)*, Iowa State College, Ames, IA.

Chapter 6

Finite element analysis

Ian D. Moore
Queens University

6.1 INTRODUCTION

Finite element analyses are now often chosen to conduct the calculations needed for specific conduit structures and embedment conditions. This chapter provides an overview of the capabilities and limitations of these analyses, illustrated with some specific examples and guidance on key issues that can be addressed in analysis and design with finite elements but does not provide instruction on undertaking analysis with any particular software package. Interested readers should seek guidance or training on the software they wish to use.

The soil-conduit interaction models presented in Chapter 5 almost all date from before the development of computer analyses of soil-conduit interaction. Each of those models is limited by the geometric and material approximations used in their development (e.g., circular conduit geometry, linear elastic soil and structural behavior, deep cover, uniform properties for the structure and/or backfill, and simplistic models of bedding and haunching support). However, many conduit structures fall outside of those limits, and so other analysis tools may be needed to perform design, assessment, and research tasks.

A key feature of finite element analysis is the discretization of the structural and soil systems into a series of small regions (the finite elements). Figure 6.1 demonstrates this for two- and three-dimensional (2D and 3D) models. Each element is defined to have its own specific individual geometry, and it may also be assigned its own material properties, allowing modeling of complex conditions such as the following:

- The different zones of soil and other materials in the vicinity of the conduit structure (see Figure 6.1a), such as layers of native soil below and beside the conduit, erosion voids or hard inclusions within the soil near the conduit, special foundation elements like loose or compacted fill or reinforced concrete used in footings, and pavement layers near the ground surface.

DOI: 10.1201/9780429162619-6

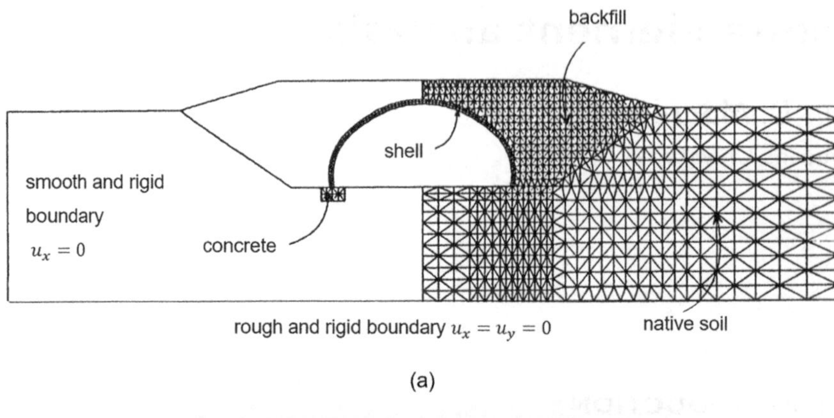

(a)

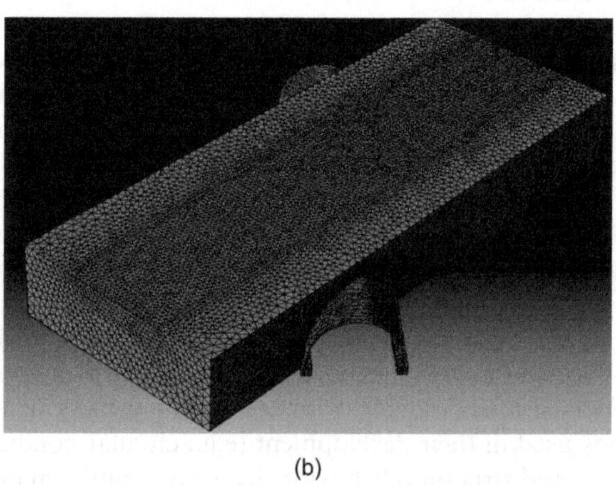

(b)

Figure 6.1 Examples of 2D and 3D conduit finite element meshes. (a) 2D mesh for 33 ft (10 m) span arch conduit (u_x = horizontal displacement, u_y = horizontal displacement). (Adapted from Taleb and Moore, 1999) (b) 3D mesh for 10 ft (3 m) span skewed conduit (Liu, 2021)

- Non-tubular (e.g., arch) shape, and/or noncircular shape, and/or non-uniform distribution of structural properties (e.g., allowing consideration of variable structural thickness within a new structure or after deterioration or varying reinforcement in a concrete structure).
- Installation-specific soil and conduit configurations as shallow burial depth, the presence of sloping ground surface, the shape of the zone of native soil that has been excavated and backfilled to accommodate the conduit (Figure 6.1a), and skewed geometry (Figure 6.1b).
- Detailed representation of the geometry of the applied loads and spatial variations in the load intensity.
- The presence of multiple conduit structures at the site, which may interact with each other.

- Changes in the material behavior of the structural and soil components of the problem as loads are applied, including material failure (e.g., yield of a steel structure or concrete reinforcement, strain hardening of the soil, shear failure of the soil).
- Changes in project geometry (e.g., explicit modeling of the placement of backfill or consideration of changes in the shape of flexible compression elements to assess buckling instability).

Each decision to incorporate additional geometrical, material, or loading complexities into the analysis requires added effort to create the model, perform the analysis, and interpret the outcomes. Critical thinking, judgment, and experience will guide decisions to optimize the choices made during analysis definition and execution while maintaining effective representation of key features that influence or control the engineering decisions.

Finite element analysis can be undertaken with many goals:

- Checking the performance of simplified design equations, either by undertaking an analysis that reflects the same approximations used when the design equation was formulated or by removing some of the approximations to explore their impact on the calculated behavior. 3D behaviors, such as the distribution of live loads through fills and within the structure, require finite element analysis to model soil stress distribution and structural response (see Figure 7.7 and McGrath et al., 2005).
- Estimating stress resultants (e.g., force, moment, or shear) and deformations (e.g., decrease in vertical diameter) where the soil, or structural geometries, or the material characteristics of the soil or conduit structure do not match the available design solutions, or when there is no simplified design solution, and the standard specifies the use of finite element analysis. For example, deep corrugated metal structures require finite element analysis per the AASHTO LRFD Bridge Design Specifications (AASHTO, 2020; called AASHTO LRFD hereafter).
- Investigating the load rating or structural capacity of existing structures, including those that feature structural or soil deterioration.
- Interpreting measurements of field or laboratory performance.
- Design a structural rehabilitation or other repair.
- Undertake parametric studies to develop new simplified design equations.

The following sections present a discussion of the use of 2D and 3D finite element analyses and specific finite element programs (e.g., CANDE (2022)). The chapter concludes with a presentation of example analyses that illustrate the performance and limitations of such finite element calculations. This chapter is focused on finite element analyses that model both the soil and structural components of the conduit, not structural analysis programs like Eriksson Pipe (Eriksson, 2021) that define the influence of the soil using equivalent pressure distributions acting on the conduit (see Section 5.2.2).

6.2 2D VERSUS 3D ANALYSES

2D conduit analyses were developed in the 1970s and 1980s (e.g., Katona et al., 1976; Duncan et al., 1980; Chang et al., 1980; Moore, 1987) to study pipe, arch, box, and other structure shapes, to interpret field measurements, to model and understand progressive placement of backfill beside the structure (e.g., staged construction), and to examine the influence of nonlinear soil and structural behavior at service loads and, in some cases, the ultimate limit state. 2D analysis features explicit modeling of the structural and soil geometries in the plane perpendicular to the conduit axis and the use of the plane strain assumption where normal and shear strains are calculated in that plane, and other strains are set to zero. 2D analysis can be highly effective at modeling the response of conduits of prismatic geometry under earth loads because earth loads can be approximated as uniform in the direction parallel to the conduit axis.

2D analyses can also be used to estimate conduit response to vehicle loads. Such analyses model load spreading and stress attenuation with depth below the ground surface in the plane parallel to the conduit span; however, it does not model variations of load in the longitudinal direction and so relies on conversion of loads of 3D geometry, Figure 6.2a, to equivalent line

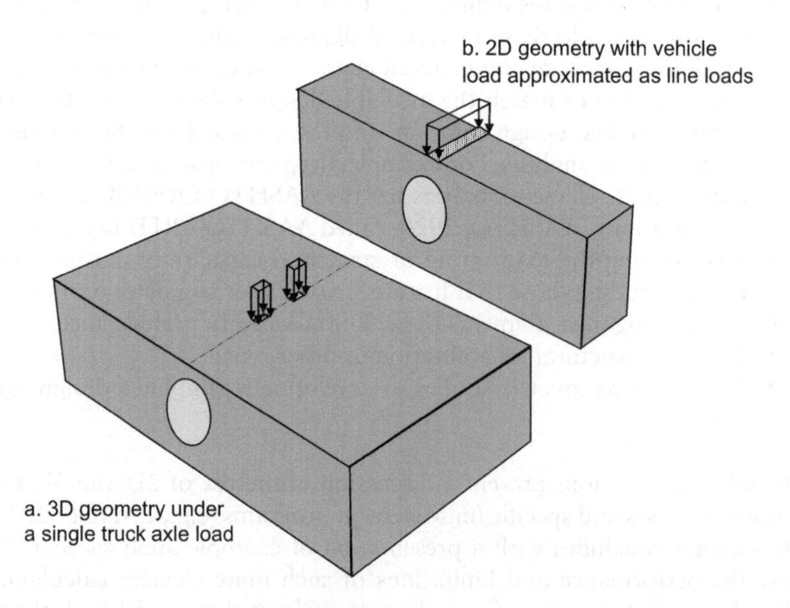

Figure 6.2 Geometrical modeling of vehicle loads.

loads, Figure 6.2b. A typical approach is to reduce the load applied at the surface of the model to account for the 3D load distribution down to the level of the top of the structure, i.e., to apply the load calculated load at the top of the conduit to the ground surface of the model. For a single wheel, each concentrated load of magnitude P across the span of the conduit can be converted to a line or strip load of magnitude $p = P/_{Fh}$, where h is the depth from ground level to the top of conduit and F is a spreading factor. Katona et al. (1976) suggested $F = 1.33$ for use in 2D finite element analysis, while load spreading with $F = 1.15$ is defined for calculations of conduit thrust using the design model in AASHTO LRFD. If multiple wheels perpendicular to the conduit span interact, such as two wheels on one axle, the load at the top of the conduit can be calculated using the equation in Section 5 and in AASHTO LRFD. Section 6.4 shows that such 2D analysis can be effective at estimating conduit response to vehicle loading but that this depends on the methods used to convert 3D loads into equivalent 2D line loads. The distribution of live load effects below the top of the conduit is not generally modeled but is not overly conservative; however, CANDE includes an algorithm to address distribution below the top of the conduit.

3D conduit analyses were developed in the 1990s, first based on a normal 2D mesh, linear elastic material behavior, and the use of Fourier analysis to model variations in the direction parallel to the conduit axis (Moore and Brachman, 1994; Moore and Taleb, 1999). Subsequently, with the availability of increased computational power, full 3D modeling has been developed (e.g., Kitane and McGrath, 2006; Elshimi et al., 2014), which permits nonlinear soil and structural behavior to be considered. 3D modeling is more complex than 2D analysis and requires significantly more computer resources. However, it permits the analysis of problems that do not have prismatic geometry, such as Figure 6.1b, and it explicitly incorporates load spreading in all directions below vehicle loads, and so can significantly improve the calculation of conduit response to vehicle loading.

Figure 6.3 presents distributions around the circumference of hoop thrust due to surface loading based on 2D and 3D calculations reported by Moore and Brachman (1994) for a 25.3 ft (7.7 m) diameter circular corrugated steel conduit tested in the field by Bakht (1980) (thrust values calculated from field measurements of strain are shown with square symbols). The finite element calculations were undertaken for two values of soil modulus: 4.4 ksi (30 MPa) with dashed lines and 11.6 ksi (80 MPa) with solid lines. Katona's surface load spreading factor $F = 1.33$ was used for the 2D calculations. Experiments on shallow buried conduits consistently show that peak live load thrust can develop near the crown, and that is reflected in the field results and the 3D analyses. The 2D analyses, however, overestimate thrusts at the springline and underestimate thrust at the crown. Generally, 3D load spreading emphasizes the response at the crown because the springlines and invert are further below the ground surface. Progressive load spreading with

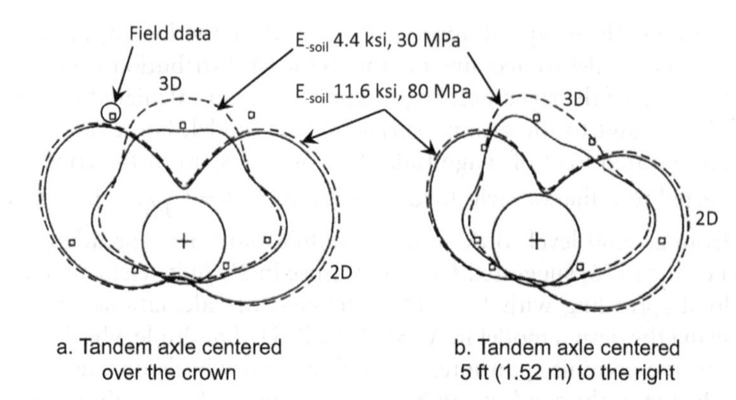

Figure 6.3 Circumferential distributions of thrust from 2D and 3D finite element analyses of response to tandem axle load (Moore and Brachman, 1994). Compared to field measurements from Bakht (1980).

depth in the direction perpendicular to the pipe axis is not captured using 2D analysis, and 2D cause conversion of applied forces to line loads is calculated based on the depth of the crown.

Moore examined results from parametric studies undertaken with 2D and 3D analyses (Appendix E from McGrath et al., 2002) and proposed spreading factors for line load calculation of $F = 11.5$ when estimating diameter decrease, $F = 3.45$ when calculating moment, and $F = 1.15$ for calculations of thrust. The higher spreading factor for assessment of deformations reflects the fact that diameter decrease depends on the response of the whole conduit cross-section, while thrust is approximately equal to radial pressure multiplied by the radius of curvature at the specific point on the conduit circumference where the thrust is to be estimated and is largely independent of the loads applied to the pipe around the rest of the pipe circumference. The performance of these different spreading factors is examined in Section 6.4.

6.3 CANDE AND OTHER COMMERCIALLY AVAILABLE SOFTWARE

Analysis can be undertaken using specialized programs developed for analyzing conduits or using general-purpose finite element programs. The specialized program most commonly used in North America is CANDE (2022) – a 2D finite element program first developed in 1976 by Katona through a contract with the Federal Highway Administration. It has subsequently been upgraded a number of times. The latest version is available as a free download on the internet.

CANDE has attractive features, including its free availability, straightforward mesh generation for conduits of standard dimensions, and ease of use. Furthermore, it is generally accepted as a conduit design tool across North

America. CANDE can also be used to analyze more general problems if a third-party mesh preprocessor is used to generate the finite element mesh. At this time, a built-in mesh generator and finer built-in meshes for CANDE are improvements that would greatly assist engineers working on conduit analysis and design.

Soil behavior in CANDE can be based on simple linear elastic analysis or based on the Duncan et al. (1980) or Selig (1988) models for soil described in Section 4.3. The Selig model and parameters are the basis for the AASHTO LRFD models for concrete and plastic conduits. Nonlinear elastic models are effective at representing soil behavior at service loads by capturing the effect of normal stresses within the soil to confine its behavior and enhance modulus, and the action of shear stresses in the soil to degrade stiffness. However, these nonlinear models cannot capture the development of plastic (permanent) strains and are thus unlikely to be effective if conduit strength is controlled by the shear strength of the backfill, if residual effects after unloading are of interest, or if response to multiple load cycles is required. In such cases, elastic-plastic soil modeling is needed to provide effective representation of the permanent strains and deformations to effectively model the soil and conduit response during unloading or reloading.

CANDE converts 3D vehicle loads to equivalent 2D live loads by using the spreading factor proposed by Katona et al. (1976), F = 1.33.

General-purpose finite element programs such as ABAQUS, ADINA, ANSYS, DIANA, LS-DYNA, and PLAXIS can be used to undertake various kinds of conduit analyses. These programs can be challenging to learn and use but provide great flexibility with respect to the details of the structural and geotechnical modeling used in the analysis. As such, they can provide solutions to challenging consulting and research problems. Advantages they offer compared to CANDE include the following:

- The ability to undertake 3D, as well as 2D, analyses.
- The ability to use a wider range of constitutive models for the soil, including elastic-plastic formulations, which permit assessment of unloading and reloading behavior and cyclic load effects.
- Some have sophisticated interaction algorithms (e.g., ABAQUS), where the conduit and the surrounding soil are not directly connected, but they interact through nodes and surfaces on the soil-conduit boundary and address adhesion, friction, and separation across the soil-structure interface. These can be used to model the Burns and Richard (1964) full-slip interface discussed in Section 5.1.
- Some have sophisticated nonlinear solvers (e.g., ABAQUS) that permit the solution of load paths involving highly nonlinear responses.
- Commercial programs generally feature detailed libraries of finite elements that permit investigation of system components that are difficult or impossible to model using CANDE, such as seams connecting corrugated steel plates or conduit end-walls (e.g., Brachman et al., 2012).

6.4 PERFORMANCE AND LIMITATIONS OF FINITE ELEMENT ANALYSIS

Many studies have been conducted comparing calculations of conduit response to measurements (e.g., Chang et al., 1980; Moore and Brachman, 1994; Moore and Taleb, 1999; Taleb and Moore, 1999; Kitane and McGrath, 2006; Elshimi et al., 2014) and each is instructive. However, the studies of Mai et al. (2014a, 2014b), which examined the performance of CANDE as well as other finite element analyses, are particularly valuable. Mai et al. (2014a) reported on the measured response of a 6 ft (1.8 m) diameter circular corrugated steel conduit under simulated axle loading, unloading, and reloading at 2 ft (0.6 m) of cover and then used those measurements of diameter change and peak thrust at or near the conduit crown to investigate the performance of 2D analyses of thrust and diameter change shown in Figures 6.4 and 6.5, respectively (Mai et al. 2014b), as follows:

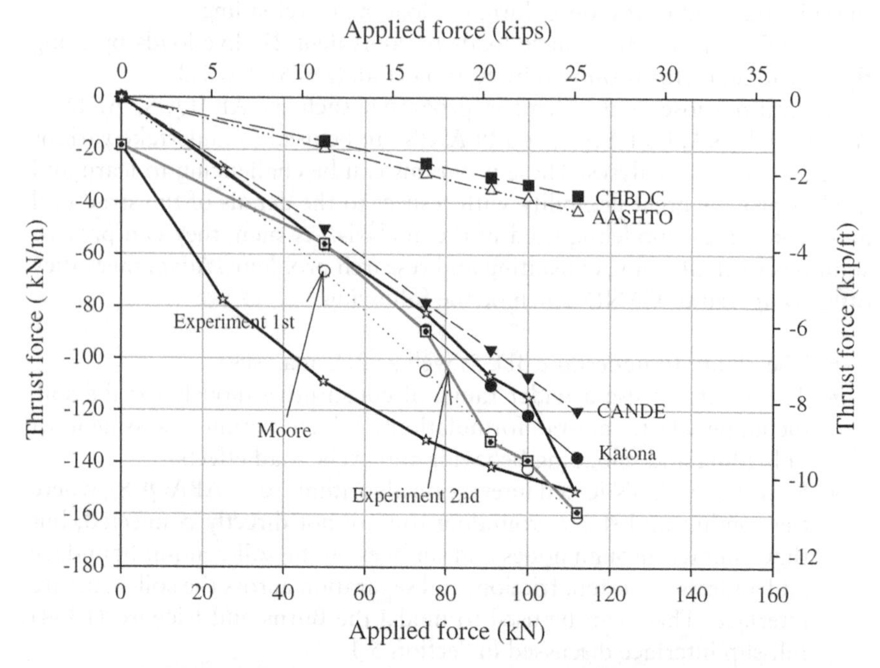

Figure 6.4 Performance assessment by Mai et al. (2014b) of 2D finite element analyses for thrust due to two single axle load cycles (CL-625 per CSA, 2019) at 2 ft (0.6 m) cover Versus Experimental Results, Mai et al. (2014a).

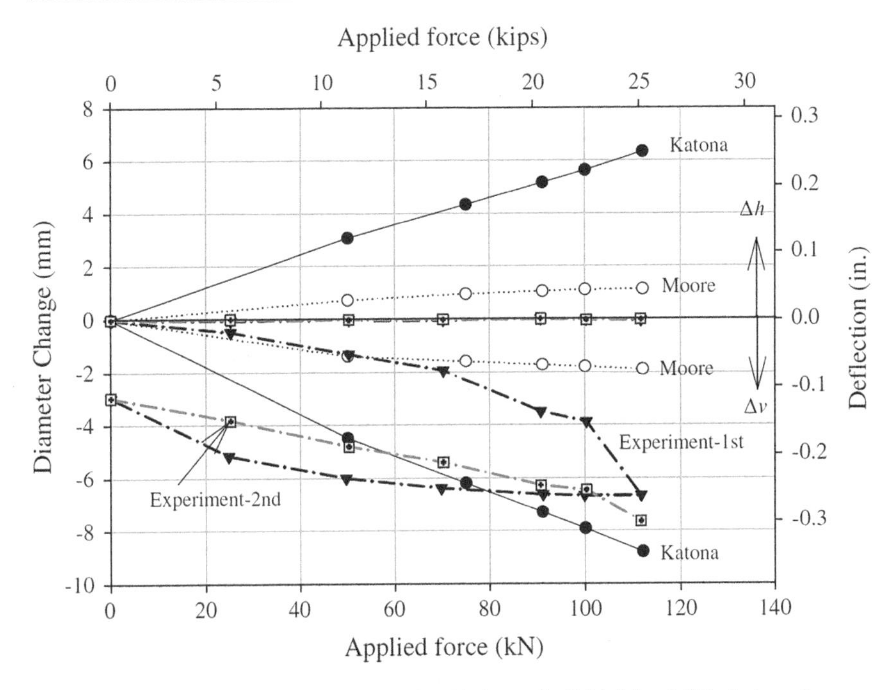

Figure 6.5 Performance assessment by Mai et al. (2014b) of 2D finite element analyses for diameter change due to two single axle load cycles (CL-625 per CSA, 2019) at 2 ft (0.6 m) of cover versus experimental (Mai et al. 2014a) with Katona or Moore load spreading factors.

- 2D finite element analysis was performed using the standard algorithms in CANDE.
- 2D finite element analysis was performed using ABAQUS based on Katona's spreading factor for the conversion of 3D axle loads into equivalent line loads.
- 2D finite element analysis was performed using ABAQUS based on Moore's spreading factors for the conversion of 3D axle loads into equivalent line loads.
- Estimated thrusts were included using the design models defined by AASHTO (2020) – an older version of AASHTO LRFD – and in the Canadian Highway Bridge Design Code (CHBDC), CSA (2019).

The comparisons shown in Figure 6.4 indicate that for this conduit, all of the 2D finite element calculations provide effective estimates of the peak thrust (though consistent with what was seen earlier in Figure 6.3, these are calculated to be near the springline instead of the crown where they were observed in the experiments) and provide much safer estimates than the two design codes. Furthermore, the thrust increases during each of the two load cycles were very similar, with thrust slightly higher at the end of the second cycle.

Current AASHTO LRFD provisions (AASHTO, 2020) provide a design factor to predict the higher thrusts experienced at the crown of metal conduits (see Section 7.6.1).

The comparisons shown in Figure 6.5 lead to more complex conclusions. There is a significant nonlinear response during the first load cycle and significant permanent deformation at the end of that cycle, i.e., once the surface load is removed. The first parts of the first and second load cycles are approximately linear, but the second half of load cycle one exhibits a significantly higher rate of deformation with load (i.e., reduced stiffness). 2D calculations based on the spreading factor of Moore exhibit decreases in vertical diameter that echoes the elastic (i.e., first) part of load cycle one and for the whole of load cycle two. The resulting estimates of the increase in horizontal diameter are also relatively close to the measured behavior. However, the 2D calculations based on the spreading factor of Katona better reflect the vertical diameter decrease observed at the highest load levels applied during the two load cycles.

Choices made when defining the geometry and material parameters for a finite element analysis should be based on an understanding of how engineering decisions depend on behavior expected in the field. As per the discussions in Chapter 4, any estimates of conduit response to incremental loads can be expected to be effective if choices of soil properties account for the dependence of soil modulus on the magnitude of the effective stresses active and the type of soil material, and how its density has been affected by its initial compaction and subsequent loading history.

The use of 3D analyses for vehicle loads can provide effective estimates of the expected magnitudes and distributions of thrust, moment, and deformation. The use of 2D representations of vehicle loads (i.e., line load equivalents) may provide effective estimates of the magnitudes of thrust for use in design but will not likely provide the correct thrust distributions. The use of 2D representations of vehicle loads can also provide effective estimates of the magnitudes and distributions of moment and deflection for use in design, provided effective scaling is used to account for the differences with which moment and deflection are influenced by 3D load spreading.

Calculations for deformations and moments due to vehicle loads may need to consider whether the load is being applied for the first time (where the soil response will be elastic-plastic) or for some subsequent load cycle (where the soil response will be approximately elastic).

Deformations due to earth loading as backfill is placed beside the structure in the field will not be estimated effectively unless the impacts of compaction loads are negligible; the direct influence of specific compaction equipment and its use will not be captured; however, bounds may be established if one of the advanced modeling techniques for estimating compaction is employed (e.g., Taleb and Moore, 1999; Elshimi and Moore, 2013).

REFERENCES

AASHTO (2020) *Standard Specification for Highway Bridges*, 17th Edition, AASHTO, Washington, DC

Bakht, B. (1980) Live Load Testing of Soil-Steel Structures, *Technical Report SRR-8-4*, Ontario Ministry of Transportation and Communication, Policy, Planning and Research Division, Toronto, CA.

Brachman, R.W.B., Elshimi, T., Mak, A., and Moore, I.D. (2012) Testing and Analysis of a Deep-corrugated Large-span Box Culvert Prior to Burial, *Journal of Bridge Engineering, ASCE*, Vol. 17, No. 1, pp. 81–88.

Burns, J.Q., and Richard, R.M. (1964) Attenuation of Stresses for Buried Circular Cylinders, *Proceedings of the Symposium of Soil Structure Interaction*, University of Arizona, Tucson, AZ, pp. 378–392.

CANDE (2022) Culvert ANalysis and DEsign, https://www.candeforculverts.com/home.html

Chang, C.S., Espinosa, J., and Selig, E.T. (1980) Computer analysis of Newtown Creek Culvert. *Journal of Geotechnical Engineering Division, ASCE*, Vol. 106(GT5), pp. 531–556.

CSA (Canadian Standards Association) (2019) *Canadian Highway Bridge Design Code*, CSA S6:19. CSA, Toronto, ON.

Elshimi, T., Brachman, R.W.I., and Moore, I.D. (2014). Effect of Truck Position and Multiple Truck Loading on Response of Long-Span Metal Culverts, *Canadian Geotechnical Journal*, Vol. 51, No. 2, pp. 196–207.

Elshimi, T., and Moore, I.D. (2013). Modeling the Effects of Backfilling and Soil Compaction Beside Shallow Buried Pipes, *Journal of Pipeline Systems Engineering and Management, ASCE*, Vol. 4, No. 4, 04013004, pp. 1–7.

Eriksson (2021), *Eriksson Pipe*, Eriksson Software Inc., Temple Terrace, FL.

Duncan, J.M., Byrne, P., Wong, K.S., and Mabry, P. (1980) *Strength, Stress-Strain and Bulk Modulus Parameters for Finite Element Analyses of Stresses and Movements in Soil Masses*, Department of Civil Engineering Report No. UCB/GT/80-01, University of California, Berkeley, CA.

Katona, M.G., Smith, J.M., Odello, R.S., and Allgood, J.R. (1976) *A Modern Approach for the Structural Design and Analysis of Buried Culverts* (FHWA Report No. FHWA-RD-77-5), Federal Highway Administration, Washington, DC.

Kitane, Y., and McGrath, T.J. (2006) Three-Dimensional Modeling of Live Loads on Culverts, *Pipelines*, ASCE.

Liu, Y. (2021) *Physical Testing and Numerical Modeling to Develop Design Equations for Corrugated Steel Culverts Under Live Loading*, PhD Thesis, Department of Civil Engineering, Queen's University, Kingston, Ontario 233pp.

Mai, V.T., Hoult, N.A., and Moore, I.D. (2014a). Effect of Deterioration on the Performance of Corrugated Steel Culverts, *Journal of Geotechnical and Geoenvironmental Engineering*, ASCE, Vol. 140, No. 2, 04013007, pp. 1–11.

Mai, V.T., Hoult, N.A., and Moore, I.D. (2014b). Performance of Two-Dimensional Analysis: Deteriorated Metal Culverts under Surface Live Load, *Tunnelling and Underground Space Technology*, Vol. 42, pp. 152–160, doi:10.1016/j.tust.2014.02.015

McGrath, T.J., Liepins, A.A., and Beaver, J.L. (2005) Live Load Distribution Widths for Reinforced Concrete Box Sections, *Transportation Research Record: Journal of the Transportation* Research *Board, CD-11-S*, Washington, DC, pp. 99–108.

McGrath, T.J., Moore, I.D., Selig, E.T., Webb, M.C., and Taleb, B. (2002) *NCHRP Report 473 Recommended Specifications for Large Span Culverts*, National Cooperative Highway Research Program, National Academy of Sciences Transportation Research Board, Washington, DC.

Moore, I.D. (1987). The Elastic Stability of Shallow Buried Tubes, *Géotechnique*, Vol. 37, No. 2, pp. 151–161.

Moore, I.D., and Brachman, R.W.I. (1994) Three-Dimensional Analysis of Flexible Circular Culverts, *Journal of Geotechnical Engineering*, Vol. 120, No. 10, pp. 1829–1844.

Moore, I.D., and Taleb, B. (1999) *Metal Culvert Response to Live Loading – Performance of Three-Dimensional Analysis, Transportation Research Record No. 1656, Underground and Other Structural Design Issues*. National Research Council, Washington, DC, pp. 37–44.

Selig, E.T. (1988) Soil Parameters for Design of Buried Pipelines. *Pipeline Infrastructure – Proceedings of the Conference*, American Society of Civil Engineers, New York, NY, pp. 99–116.

Taleb, B., and Moore, I.D. (1999) *Metal Culvert Response to Earth Loading – Performance of Two-Dimensional Analysis, Transportation Research Record No. 1656, Underground and Other Structural Design Issues*, National Research Council, Washington, DC, pp. 25–36.

Chapter 7

Structural design of soil-conduit systems

Structural design procedures for gravity flow conduits did not result from a single stream of research. Rather, it developed product by product, mostly independently of other products and sometimes drawing on prior knowledge. Some of this historical work was presented in Chapter 2. The result of this slow and diverse development is a wide variety of design methods, some empirical based on field experience and others more rigorous. This diversity often leads engineers to think all products are different – the most obvious difference being rigid versus flexible. However, as demonstrated in Chapter 5, conduit behavior is a function of the ratios S_B and S_H, and conduit products fall into different parts of a continuum based on those parameters. One goal would be to develop a universal design method that could address all conduits. This would make it easier to compare products, such as deciding which product is most appropriate for a project, but meeting this goal is unlikely in the short term. The simplifications available by focusing on a single type of product make the existing methods easy to work with, while a single comprehensive method may be more complicated than warranted given that so much of long-term performance is controlled by the installation process, which is subject to variation.

Analysis to determine the forces in a buried conduit was presented in Chapters 5 and 6. This chapter addresses the design of conduits to carry those forces. The key aspects of design for each type of product in Article 12 of the AASHTO LRFD Bridge Design Specifications (AASHTO 2020 – referred to as AASHTO LRFD) and how they are addressed by the design equations are explained, but this is not a rigorous equation-by-equation tutorial. Engineers must always use the appropriate design code, such as AASHTO LRFD, as the primary basis for calculations.

7.1 LOAD AND RESISTANCE FACTOR DESIGN

Load and resistance factor design (LRFD) is a probabilistic design approach intended to assure the safety of structures by assigning factors based on known variability of loads (load factors, γ, in AASHTO LRFD) and known

DOI: 10.1201/9780429162619-7

variability of material strengths (resistance factors, ϕ, in AASHTO LRFD). In LRFD design, the likelihood of reaching a limit state can be evaluated in a rational fashion that allows design for uniform safety across all the elements of a structure. However, assessing the safety of soil-conduit systems is challenging and has not been rigorously addressed at this time. The wide variety of design methods used for gravity flow conduits and the proper treatment of soil contribution to performance make it difficult to complete a probabilistic assessment. Prescriptive and empirical methods contribute to the difficulty. Thus, most of the conduit design methods presented here have the appearance of LRFD but are largely a reimplementation of load factor design procedures with the addition of resistance factors that are mostly set to 1.0. Nevertheless, all practitioners should understand the elements of LRFD and the AASHTO approach to safety. Galambos (1981) presents the basic elements, and AASHTO LRFD Article 1 provides background related to bridges.

Basis of LRFD – Safety of structures and structural elements is based on keeping the material capacity, called the resistance, *R*, greater than the applied loads, *Q*. Since the resistance and loads are both variables, usually assumed to be normally distributed, there is always some probability of failure in the region where the two probability density functions overlap (Figure 7.1) – that is, where $R < Q$.

Safety of the structure or element can be defined in terms of the ratio of R/Q, which has a probability density function just as the individual variables *R* and *Q*. Safety is provided by identifying limit states and setting load and resistance factors to provide a suitably low probability of failure. In a plot of the natural logarithm of the probability density function, $\ln(R/Q)$, the region less than zero represents the probability of failure (Figure 7.2). The risk of failure is reduced as the mean value of $\ln(R/Q)$ increases above zero. LRFD uses the reliability index, β, which represents the number of standard deviations, $\sigma_{\ln(R/Q)}$, between 0 and the mean of $\ln(R/Q)$. The larger the reliability index, the lower the probability of failure.

AASHTO LRFD Article C.1.3.2.1 states that the target reliability index for the Strength 1 limit state, which is the AASHTO governing strength limit state for conduits, is 3.5, which corresponds to a probability of exceedance of 2 x

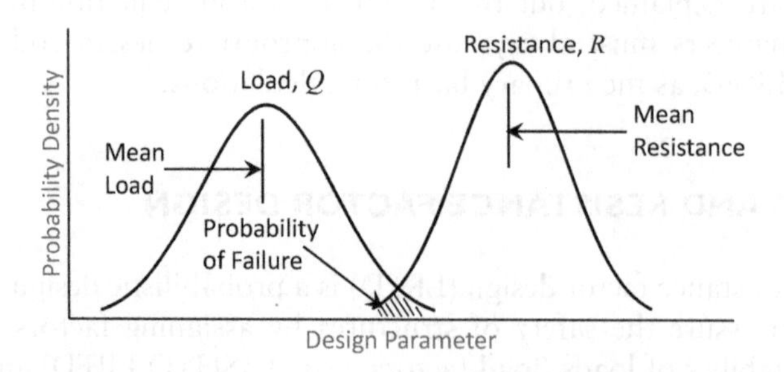

Figure 7.1 **Probabilistic relationship of load and resistance.**

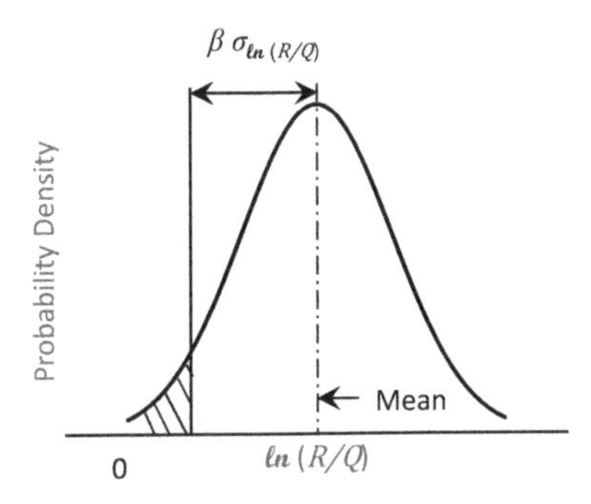

Figure 7.2 Definition of the reliability index, β.

10^{-4} in 75 years and an annual probability of exceedance of 2.7 x 10^{-6} with a target reliability index of 4.6. Note that β is based on ln (R/Q), and a change of 1 represents a significant change in safety. The probability of failure for a reliability index of 3 is approximately 10^{-3}, while the probability of failure for a reliability index of 4 is substantially lower, approximately 3 x 10^{-5}.

7.2 LIMIT STATES

A limit state is a condition of a structure beyond which it no longer fulfills the relevant design criteria. AASHTO LRFD identifies four classifications of limit states: service, fatigue and fracture, strength, and extreme event strength. Service limit states are "restrictions on stress, deformation, and crack width under regular service conditions" (AASHTO LRFD Article 1.3.2.2). Fatigue and fracture are not addressed in the routine design of conduits but could be considered when setting stress limits. Strength limit states ensure that strength and stability are provided to resist conditions that exceed the service limit but that a structure may experience during its service life. Strength limit states make up the largest portion of conduit design calculations. Designing for extreme events ensures the survival of structures during earthquakes, floods, and other rare events. Such events have not historically been addressed in routine conduit design; however, with millions of conduits passing under roadways that are vital to both routine transportation and emergency services, as well as the increasing frequency of catastrophic storms, more attention is being paid to ensuring conduit resilience, both structural and hydraulic, during extreme events.

7.3 LOAD AND RESISTANCE FACTORS

Research has not yet been undertaken to determine the probabilistic profile of loads on, and resistance of, buried conduit installations. Thus, AASHTO LRFD load factors for earth loads on conduits are mostly values that were set prior to the introduction of the AASHTO LRFD specifications and vary from product to product while, with the exception of concrete conduits, resistance factors are mostly set to 1.0. AASHTO LRFD load factors are discussed here.

Load Factors, γ – AASHTO LRFD considers a range of limit states for conduits. These conditions and their assigned load factors are provided in the Strength I load combination in AASHTO LRFD Table 3.4.1-1 and for permanent loads (e.g., vertical earth pressure) in Table 3.4.1-2:

Structure weight (DC) is the weight of the conduit. This is often not considered in the design of flexible conduits as the structure weight is small relative to earth and live loads. It should be considered for heavy conduits such as concrete. A maximum load factor of 1.25 is specified when the structure weight increases the structural demand (i.e., stress, strain, force, or moment) at a location under consideration and 0.9 if the weight of the structure decreases the structural demand under consideration. The maximum/minimum load factors (henceforth listed in max/min format) are an important aspect of design, as some load conditions will increase demand at one location while reducing it at another.

Wearing surface (DW) represents the weight of the wearing surface. This load condition has a wider range of load factors of 1.5/0.65 to account for variation in pavement depth due to application of overlays and wear over time. For buried structures, pavement is typically a small load relative to earth and live loads. Pavement thickness is commonly addressed by adding it to the depth of fill and assigning it the backfill unit weight.

Horizontal earth pressure (EH) is the lateral pressure on a conduit. Flexible conduit design considers horizontal earth pressure as part of the resistance and not as a load. Box conduits and other conduits with vertical sides consider 'at rest' horizontal pressure and AASHTO states that the lateral pressure coefficient need not be taken greater than 0.5 (2023 ballot item). AASHTO specifies load factors of 1.35/0.9 for lateral earth pressure, but many box conduits have been designed with maximum/minimum lateral pressure coefficients of 0.50/0.25 with a single load factor. Either method achieves the intent of considering a range of conditions.

Vertical earth pressure (EV) is the primary load on buried conduits with more than a few feet of earth fill. The varied development of conduit design methods has led to a variety of approaches to this condition, some due to soil-conduit interaction effects and some empirical. AASHTO LRFD provides separate load factors as a function of the structure type:

- Rigid buried structures include concrete pipes, box conduits, and arches. The AASHTO LRFD load factors are 1.30/0.90. An exception to this is nonrectangular concrete conduits designed by the indirect method where the load factor for the strength limit state is 1.5 for some conduits and 1.25 for others for both earth and live loads (see Section 7.5.2.1 "Indirect Design").

- Metal box conduits, structural plate conduits with deep corrugations, and fiberglass conduits have load factors of 1.50/0.90. This category is for structures designed for flexural resistance and does not apply to thrust resistance.
- Plastic conduits, which include those manufactured from PE, PVC, PP, and fiberglass, have vertical earth load factors of 1.95/0.90. The horizontal earth pressure is derived from soil-conduit interaction and is not considered as a separate load.

Two types of metal conduits do not have required load factors listed in Table 3.4.1.2:

- Metal pipe, pipe arch and arch structures, and steel-reinforced thermoplastic conduits are designed for thrust under AASHTO LRFD Article 12.7.2.2, which in turn refers to LRFD Article 12.12.3.4 on thermoplastic conduits to calculate the thrust force. This implies that the load factors for these structures should be the same as for thermoplastic conduits 1.95/0.90.
- Long-span structural plate structures are designed prescriptively based on guidance in the specifications. For example, stiffeners are required, but there are no requirements provided for the stiffeners to "increase the moment of inertia of the section to that required for design" (AASHTO LRFD Article 12.8.3.5.2). This is discussed in more detail in Section 7.6.2.

Vehicular live loads (LL) – The factor for vehicular live loads, called live loads hereafter, and dynamic allowance (impact) loadings resulting from live loads is 1.75.

Water load (WA) The load factor for water is taken as 1.0. Water levels should be specified to produce appropriately conservative designs.

For circular conduits, hydrostatic pressure produces a compressive thrust, which is usually only significant for thin-walled conduits, such as those with corrugated profiles. In noncircular conduits, the vertical and horizontal external pressures are increased by hydrostatic pressure, increasing bending moments.

Design for hydrostatic pressure requires using the bulk (wet) unit weight for soil above the groundwater table and the buoyant unit weight below the groundwater table, as well as the hydrostatic pressure. Variation in the groundwater table should be evaluated to find an appropriate governing condition.

Applying load factors – Design with a range of load factors is easily incorporated into some computer programs, such as concrete box conduit programs, which apply loads to a frame model. Each load condition can be analyzed separately with the appropriate load factor. In finite element analysis, which is used for the design of several types of large-span conduits, it is difficult to isolate each load condition to apply the appropriate factors, particularly for vertical and horizontal earth loads. If an analysis is conducted for factored earth loads, increasing the soil unit weight increases the vertical stress in the sidefill, which in turn increases the lateral pressure and provides support to the conduit. Some designers conduct analyses for service loads and then apply load factors to the resulting forces. This does not mimic a true limit state but is sufficiently conservative for design. This subject was discussed in Chapter 6.

Load modifiers – AASHTO LRFD Article 1.3 further addresses safety with the use of load modifiers as a function of ductility, η_D, redundancy, η_R, and operational importance, η_I. Each load modifier is set to 0.95, 1.0, or 1.05 as guided by the specifications. For example, if a structure or structural element is nonductile, nonredundant, or critically important, the load modifier is set to 1.05, while designs with conventional ductility or with enhanced ductility would warrant a load modifier of 1.0 and 0.95, respectively. AASHTO LRFD, Article 12.5.4 states that buried structures are considered nonredundant under earth loads (η_R = 1.05) and redundant under live and dynamic allowance loads. The expected level of redundancy under live loads, i.e., whether the load modifier could be set to 1.0 or 0.95, is not clear. Given the ability of conduits to mobilize soil support or transfer load to lower stressed sections, setting η_D = 1.0 seems reasonable.

Resistance factors – Resistance factors for buried conduits are provided in AASHTO LRFD Article 12.2.2.1. As noted, for the most part, resistance factors for buried conduits have not been calibrated to determine the statistics on variability and are thus set to 1.0, pending the results of any future calibration efforts. Variations from this general treatment include the following:

- Seams in joints of metal conduits have resistance factors set to 0.67, which makes the effective multiplier R/Q = 3, the traditional load factor for joints.
- Concrete conduits are assigned resistance factors as a function of the structure type and the force being addressed. Values are provided for flexure, shear, and radial tension, ranging from 0.82 to 1.0. This is similar to the approach used in the LRFD design of concrete bridges and buildings.

7.4 VEHICULAR (LIVE) LOADS

The primary vehicular loading in AASHTO LRFD is the HL-93 load. This load is made up of a design truck (Figure 7.3), which includes two 32 kip (142 kN) axles spaced at 14 ft to 30 ft (4.3 m to 9.1 m) and a design tandem, which consists of two 25 kip (110 kN) axles spaced 4 ft (1.2) m center

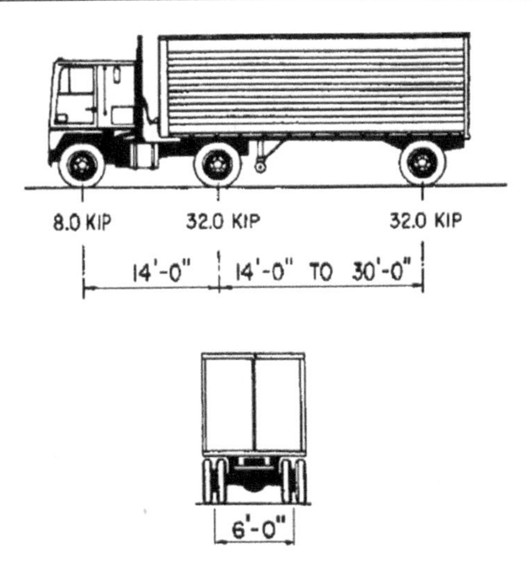

Figure 7.3 Configuration of AASHTO design truck. (From AASHTO LRFD Bridge Design Specifications, 2020, published by the American Association of State Highway and Transportation Officials, Washington, DC. Used with permission)

to center. The wheels on an axle are spaced 6 ft (1.8 m) center to center for both the design truck and the design tandem. The typical design lane width is 12 ft (3.7 m), although there are exceptions. Although not required, some computer programs model the design tandem as a design truck with the rear axle replaced with the design tandem.

AASHTO also specifies a lane load of 0.64 kip per lineal foot (9.3 kN/m) to be spread over a 10 ft (3 m) width and applied simultaneously with the design truck or design tandem load. This load represents lighter vehicles that might be on a bridge at the same time as the design truck. Applying this load in addition to the HL-93 load is unnecessary for most smaller span conduits as the length of the design truck is 28 to 44 ft (8.5 to 13.5 m), and there can be no other traffic over the conduit. If an agency decides it should be included, the lane load is approximately the weight of 6 in. (150 mm) of soil, which is not a significant load for a buried conduit.

AASHTO LRFD specifies that truck and lane loads be placed within a 12 ft lane to create the maximum force effect. In the case of passing trucks, wheel pairs from vehicles in adjacent lanes could be 4 ft (1.2 m) apart, center to center. However, for conduits, the multiple presence factor (see Section 7.4.1) is reduced from 1.2 for a single truck to 1.0 for passing trucks, and the single truck condition still controls the design.

The basic distribution of live loads is presented in AASHTO LRFD Article 3.6.1.2.6, which provides detailed equations for calculating live load distribution and is discussed in the next section. Further modifications of the live load for metal and thermoplastic conduits are discussed in Section 7.4.2.

7.4.1 Live load modifiers

Multiple presence factor (AASHTO LRFD Article 3.6.1.1.2) – A multiple presence factor (m) is employed to consider likely vehicle loads when trucks are simultaneously in one, two, three, or more adjacent lanes. Studies show that overloaded trucks are so common on roadways that the multiple presence factor for a single loaded lane is set to 1.2, increasing the primary axle load on the design truck to 38.4 k (170 kN). The probability of simultaneous adjacent heavy trucks drops with each added lane, and thus m = 1.0, 0.85, and 0.65 for trucks in two, three, or more than three lanes. AASHTO LRFD Article 3.6.1.2.6a states that conduits shall be analyzed for a single loaded lane, with m = 1.2 when the vehicle is moving parallel to the span (Figure 7.4a) because a single truck with m = 1.2 is the governing load condition over passing trucks with m = 1.0 or less. Vehicles moving perpendicular to the span direction must be evaluated for multiple loaded lanes, as axles on adjacent vehicles will load the conduit span at the same time (Figure 7.4b). AASHTO allows the use of lower values of the multiple presence factor on bridges where average daily truck traffic is 1,000 vehicles or less.

Dynamic (impact) load allowance (AASHTO LRFD Article 3.6.2.2) – AASHTO LRFD requires consideration of dynamic loads on buried conduits. This load, usually referred to as the impact load, accounts for uneven road surfaces and other anomalies that can cause vehicles to rise and fall, creating additional load. The dynamic load allowance (IM) is expressed as a factor to be multiplied by the vehicle weight with a value of 1.33 if the top of the conduit is at the ground surface (flat-topped concrete conduits can have zero cover) decreasing linearly to 1.0 when the top of the conduit is at a depth of 8 ft (2.4 m) or greater.

$$IM = 1.33 - 0.33\left(H / k_u\right) \geq 1.0, \tag{7.1}$$

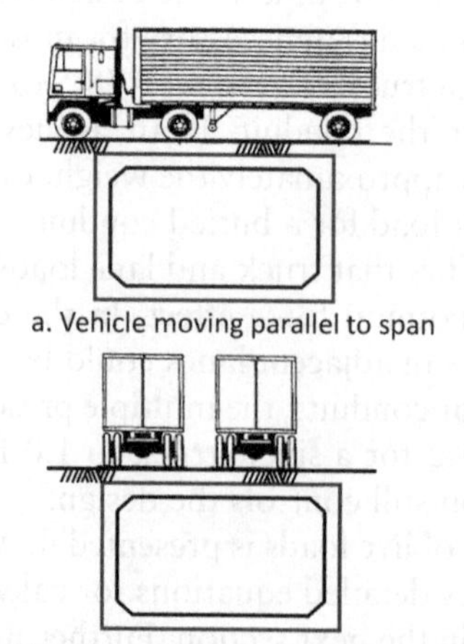

a. Vehicle moving parallel to span
b. Vehicle moving perpendicular to span

Figure 7.4 Traffic directions relative to span direction.

where

IM = Dynamic load allowance
H = Depth of fill over top of conduit, ft, m
k_u = Unit correction factor, 8 ft-in US units, 2.4 m in SI units (k_u is used to convert several equations – the values vary depending on the conversions required)

Figure 7.5 shows the factored earth and live loads versus depth of fill for a single loaded lane with the design truck traveling parallel to the conduit span increased to account for multiple presence and impact loading. The earth load is calculated with a soil unit weight of 120 pcf (18.8 kN/m^3) and a vertical arching factor, VAF = 1.0. The figure demonstrates that live loads are the primary load at shallow depths, but they attenuate quickly. The contributions from live and earth loads are approximately equal at 5 ft (1.5 m), and live load is less than 10% of the total load at a depth of about 12 ft (3.7 m), which is approximately the depth at which HL-93 loads can be dropped as a load condition (Note – this latter provision was adopted after printing of AASHTO LRFD 9th Edition). When designing conduits for very shallow fills where the live load changes rapidly for small changes in depth of cover, it is prudent to consider the possibility of installation at less than the design depth of fill. For example, if the depth of fill is 15 in. (380 mm), the factored earth plus design truck load factored pressure is 6.0 ksf (0.29 MPa), but if

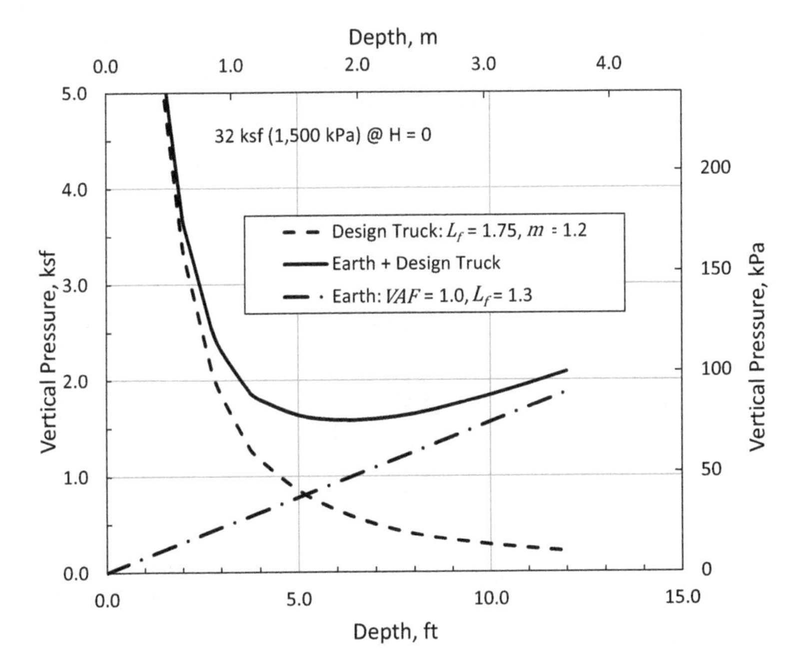

Figure 7.5 Factored earth and live loads vs depth of fill.

the actual installation depth is 12 in. (300 mm), the factored pressure increases to 7.9 ksf (0.37 MPa), a 30% increase in load.

7.4.2 Distribution of live loads to conduits

Attenuation of live loads through soil was introduced in Section 5.3. Once the live load reaches a buried conduit, the interaction of the conduit and soil results in different load paths as a function of the geometric and stiffness properties of the conduits and soil, i.e., the stiffness ratios S_B and S_H. The current live load distributions in AASHTO, which are presented here, were developed by Petersen et al. (2010). If designing in accordance with AASHTO, a careful reading of the AASHTO LRFD provisions is required, as the Petersen method introduces new concepts.

Circular and rectangular concrete conduits with high hoop and bending stiffness resist live loads primarily by developing bending moments. Concrete arches and arch top three-sided conduits can carry significant amounts of live load through thrust forces, as well as bending moments. Metal conduits and fiberglass conduits, with high hoop stiffness but low bending stiffness, resist live loads primarily by developing thrust forces. Thermoplastic conduits, corrugated PE and PP in particular, have low hoop and bending stiffness, allowing the conduits to compress circumferentially away from the load, some of which transfers to the soil embedment. These three behaviors are illustrated in Figure 7.6.

Flat-topped and arched concrete conduits with depth of fill less than 2 ft (0.61 m) – Flat-topped and arched concrete conduits with less than 2 ft of fill are treated as bridge decks and have different load distribution procedures than other conduits. Concrete bridge decks are structurally stiff, both parallel and transverse to the span direction. The transverse stiffness distributes load perpendicular to the span to a width wider than the distribution through soil presented in Section 5.3. Equation 7.2 accounts for this extra

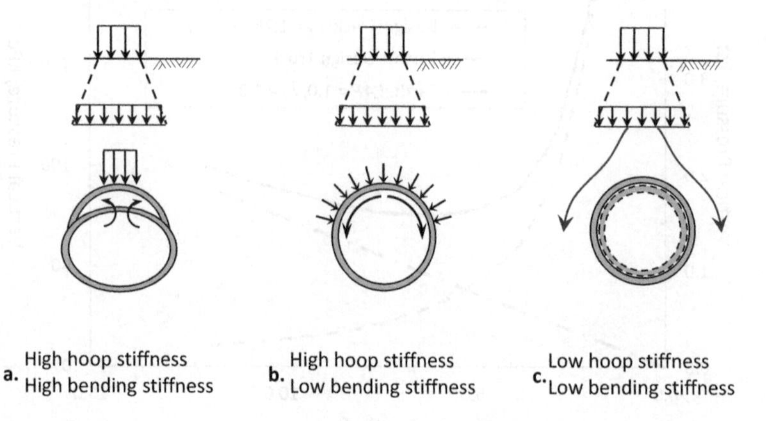

Figure 7.6 Load path for live loads in buried conduits.

stiffness in AASHTO LRFD. The equation is applicable for depths of cover up to and including 2 ft (0.61 m).

$$E = k_u + w_t + 0.06\ S_i \leq 6\,ft,\ 2.1\,m, \tag{7.2}$$

where

E = strip width perpendicular to span that a wheel load may be distributed over, ft, m
w_t = width of wheel footprint perpendicular to span, ft, m
S_i = inside span of conduit, ft, m
k_u = Unit correction factor, 2.3 ft-in US units, 0.70 m in SI units

The limitation on E is suggested in light of the increasing span of conduits in current practice. It limits the benefit of the term $0.06(S_i)$ to spans up to about 33 ft (10 m).

Equation 7.2 is used in conjunction with 2D frame models; thus, longitudinal distribution is the same for all force resultants, moment (M), thrust (N), and shear (V). However, in actual conduits, the distribution widths vary for each type of force, increasing successively for shear, positive moment (M+), and negative moment (M-). McGrath et al. (2005) documented this distribution, which is shown in Figure 7.7. This alternate distribution is

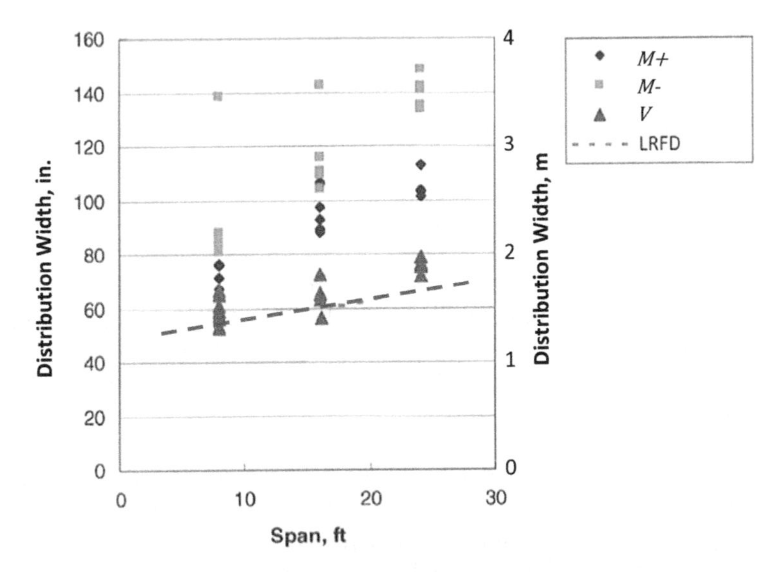

Figure 7.7 Distribution width of design forces for wheel loads on box conduits with zero cover. (Adapted from McGrath et al., 2005, Figure 5a, p. 106. Copyright National Academy of Sciences. Reproduced with permission of the Transportation Research Board)

typically not significant but can sometimes be applied to assist in load rating of conduits. For example, if the design moment at the critical shear section is reduced to reflect the distribution in Figure 7.7 then the shear capacity will be increased as shown in Section 7.5.3.

The distribution width parallel to the span for depths of cover from 0 to 2 ft (0 to 0.6 m) is the same as used for conduits with more than 2ft (0.6 m) of cover, which is the tire dimension in the direction parallel to the span plus the live load distribution factor times the depth of fill, i.e., $w_t + LLDF\ (H)$.

Live load distribution for rectangular sections with more than 2 ft of cover and nonrectangular concrete, metal, and plastic conduits with one foot or more cover – The approach to live load distribution for these conduits was presented in Section 5.3.

- AASHTO LRFD Article 3.6.1.2.6 presents the equations for distribution in accordance with Section 5.3 but adds the width 0.06 times the span (0.6 S_i) to the distribution width perpendicular to the span. As noted earlier, it is suggested to limit this increase to 2 ft (1.2 m).
- For nonrectangular concrete conduits, LLDF = 1.15 for diameters/spans less than 2 ft (0.6 m) and 1.75 for diameters/spans 8 ft (2.4 m) and greater. The LLDF is calculated by linear interpolation for intermediate sizes. Use of LLDF greater than 1.15 is due to the higher longitudinal stiffness of these conduits relative to metal or plastic conduits.
- For metal and plastic conduits, LLDF = 1.15 for all diameters/spans.
- Live loads must be increased by the appropriate dynamic load allowance, load factor, and multiple presence factor.
- The equations in Article 5.3 are used to determine the vertical load on the top of the conduit, which provides the springline thrust. In flexible conduits at low depths of fill, crown thrust forces are higher than at the springline because of the low bending stiffness relative to concrete conduits. Computation of these higher thrust forces is addressed in Sections 7.6.1 and 7.7.2.
- Live load distribution for conduits with less than 1 ft (0.3 m) of fill or less than the minimum allowable depth of fill discussed next must be analyzed by more comprehensive methods than those included in AASHTO LRFD. This usually requires 3D finite element analysis (see Chapter 6).

Approaching wheel load – The vertical stress in the soil due to a wheel load is accompanied by an increased lateral soil pressure on the conduit. As a vehicle approaches a box conduit, this increased lateral soil stress becomes a load applied to the conduit. Figure 7.8 shows the AASHTO LRFD pressure distribution to represent this load. This loading condition is only applied to flat-topped conduits with 2 ft (0.6 m) fill or less. This pressure distribution is a recent change to AASHTO LRFD, which formerly treated box conduits

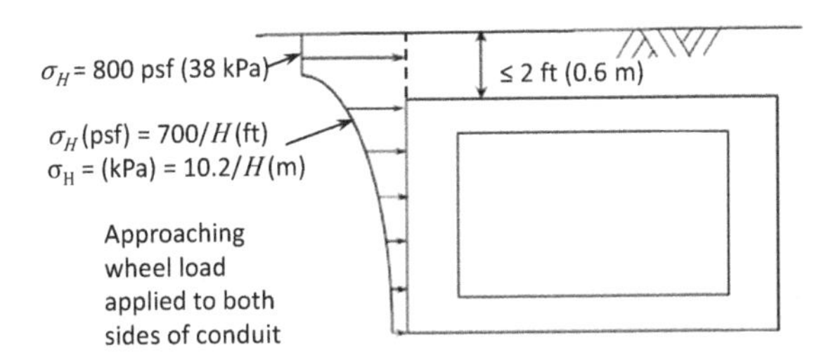

Figure 7.8 Approaching wheel load pressure on box conduit.

as if they were retaining walls, where the load is significant and could result in rotation of the wall. Since box conduits have soil on both sides of the conduit, a more realistic and lower magnitude pressure is applied.

Minimum depth of cover – The factored live load from a design truck with zero cover is about 32 ksf (1,500 kPa). Rectangular concrete conduits are designed like bridge decks, but nonrectangular conduits require 1 ft (0.3 m) or more cover to allow attenuation of the load prior to reaching the conduit unless a special design is prepared. Some conduits can carry live loads at lower depths for single or multiple load cycles, but design agencies consider conduit lifetimes to be 50 years or more, with thousands or perhaps millions of load cycles. Minimum depths of fill are set to assure good long-term performance under these actual conditions and to protect the pavement from hard spots in the case of rigid conduits or soft spots in the case of flexible conduits. Conduits with curved tops, i.e., pipes and arches, generally have minimum depths of fill of 1 ft (0.3 m) or greater. The minimum depth typically increases with diameter or span and often increases with decreasing bending stiffness.

In some cases, the minimum depth of fill is based on environmental conditions. During a test of 60 in. (1,500 mm) diameter corrugated polyethylene conduit in a cold climate (McGrath et al., 2005), the diameter reduction in winter due to thermal contraction was significant and resulted in a dip in the pavement. This is not typical of other types of conduits because the coefficient of thermal contraction of polyethylene and polypropylene is about ten times greater than that of steel. This observation led to setting the minimum depth of fill for thermoplastic conduits in AASHTO LRFD to $D_i/2 \geq 2$ ft (600 mm) under paved roadways to minimize pavement dips caused by thermal contraction in cold weather.

7.5 STRUCTURAL DESIGN OF CONCRETE CONDUITS

Concrete conduits generally are classified as box conduits or pipe conduits. This Section presents information on analysis and design of both types. The calculations can also be completed using available software: ACPA/CPPA (2023), Eriksson (2021a), and Eriksson (2021b).

7.5.1 Concrete box conduits

Concrete box conduits are rectangular with three or four sides, as shown in Figure 7.9. Four-sided boxes can be cast-in-place or precast, while three-sided boxes are only precast. Haunches are used in most box conduits, but the size and angle vary. Precast box conduits manufactured in accordance with ASTM standards have 45° haunches (Figure 7.9a) the same thickness as the side walls. Some box conduits have haunches that extend out into the top slab (Figure 7.9b). Three-sided sections can be used where the design calls for a natural stream bed or for other reasons. Footings for three-sided sections are typically cast-in-place but can be precast. Three-sided box installations for streams must be designed to avoid scour during storms that could undermine the footings. There are also three-sided boxes with arched tops and vertical sides, which are discussed later, as they are typically analyzed using finite element methods.

7.5.1.1 Structure modeling

As noted, most box conduit design is conducted by analyzing a four-member frame subjected to various applied pressures (Figure 7.10). In such a model, all applied loads must be balanced vertically, horizontally, and rotationally such that there are no external reactions at the supports.

It is important for the haunch model to provide the proper distribution of moments. A haunch stiffens the corner of the box, increasing the negative moment (tension on the outside surface) at the corner and decreasing the positive moment at midspan. While the negative moments are increased, the presence of the haunch moves the critical design location for a negative moment away from the corner to the tip of the haunch (intersection of the haunch with the wall or slab), and thus the negative reinforcement is less than for a design without haunches. The critical location for the negative moment in the slabs of a box conduit with 45° haunches is often near the inflection point (point of zero moment). Some computer programs use numerical techniques to model the haunch stiffness. A simple approximation that provides reasonable results is shown in Figure 7.11, where T_T and T_s represent the top slab and

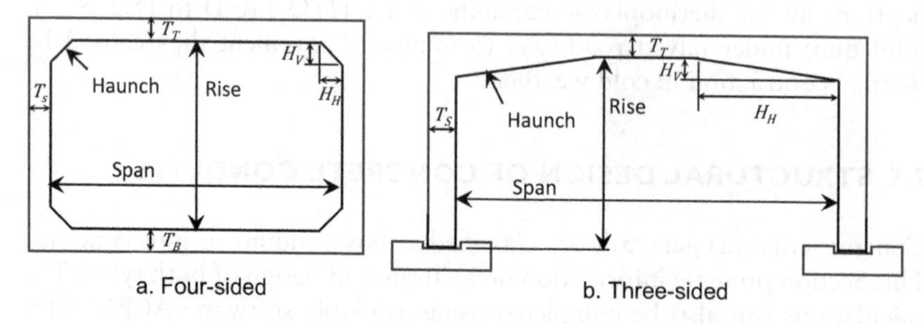

Figure 7.9 Box conduit notation. (a) Four-sided and (b) three-sided.

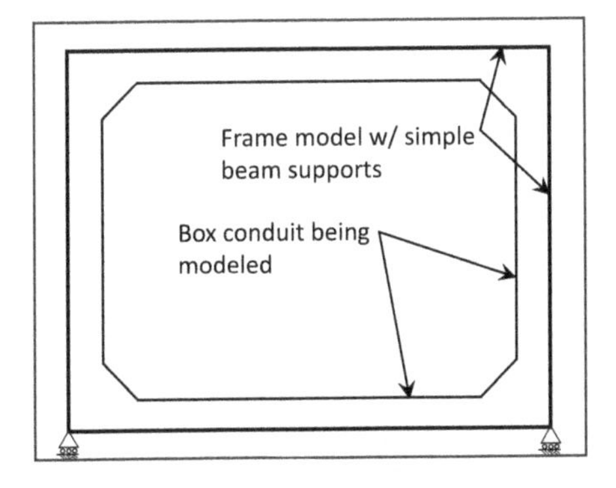

Figure 7.10 Structural frame model for box conduit.

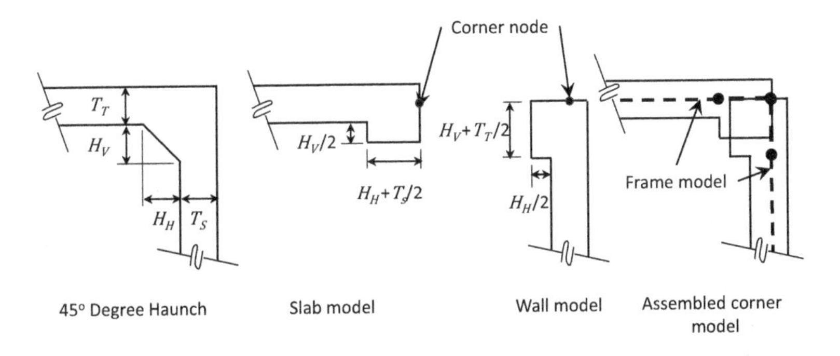

Figure 7.11 Simplified haunch model for frame analysis of box conduit.

sidewall thickness, respectively. Finite element analysis is not generally used for routine box conduit analysis and design.

7.5.1.2 Structural analysis

Concrete box conduits are typically analyzed for the following load conditions:

- Structure weight
- Vertical earth load
- Horizontal earth load
- Internal water load
- Live load – vertical loads and approaching wheel loads

Vertical earth load – Most box conduit design is completed by assuming earth load pressures (γ_sH) are uniformly distributed over the top and bottom slab. For embankment loadings, the earth load is modified by a vertical arching factor (*VAF*, called F_e in AASHTO LRFD):

$$W_E = VAF\gamma_s H B_c, \tag{7.3}$$

where

W_E = earth load on a unit length of conduit, lb/ft, kN/m along the span
VAF = vertical arching factor,
$= 1 + 0.2\ H/B_C \leq VAF_{Max}$
VAF_{Max} = 1.15 for boxes with compacted sidefill
= 1.40 for boxes with uncompacted sidefill
γ_s = unit weight of soil, lb/ft^3, kN/m^3
H = depth of fill over conduit, ft, m
B_C = outside span of box conduit, ft, m

For box conduits in trench installations, the Marston load theory, which accounts for the reduction in vertical load due to soil shear stresses at the trench wall/backfill interface, may be used. This theory is explained in Section 7.5.2.1.

If the method of analysis considers soil-conduit interaction, such as in finite element analysis, the top and bottom slabs will deflect under the earth load, resulting in shear stresses in the soil that will transfer some of the load from midspan to the corners of the section (arching). In this case, both the positive and negative moments will be reduced. If the analysis also considers cracking (cracking that does not exceed the specified service limit state) in the concrete, the deflection and arching effect will be further increased. Uncertainty regarding the soil properties generally limits use of this modified load distribution in routine design, but if soil properties are known, this behavior can be analyzed with finite elements or by applying the loads through springs with stiffnesses representing the modulus of subgrade reaction. Such analyses can be useful to analyze existing box conduit installations, such as for load rating purposes or to investigate capacity to allow the fill height to be increased over an existing box conduit. Values for the modulus of subgrade reaction should be selected by the engineer based on knowledge of the bedding and subgrade conditions.

Horizontal earth load – Horizontal earth load is generally applied as a lateral pressure coefficient, typically the coefficient for lateral earth pressure at rest, K_o, times the free field vertical stress, i.e., $K_o \gamma_s H$. Variations in lateral load are addressed by varying K_o (0.25 to 0.50 was used by ASTM for years) or varying the applied load factor. The former is more conservative, but AASHTO specifies the latter which is consistent with the overall approach to LRFD. Both methods are in common use.

Water load – water inside a box conduit creates outward lateral pressure on the sides of the box and can cause a modest increase in positive moments.

This is a minor load that is often ignored. If the external water table rises above the sides of a box conduit, the hydrostatic pressures should be considered.

Live loads – Live loads are distributed in accordance with the procedures discussed in Section 5.3 and Section 7.4 and then reduced to the load on the unit length of conduit being modeled, e.g., the area and moment of inertia inputs typically represent a 1 in. (25 mm), 1 ft (300 mm), or 1 m (3.3 ft) length of the conduit. As noted earlier, when using the four-member frame model (Figure 7.10), the reaction to the live load must be applied to the bottom slab in a way that produces no reaction. When the wheel is near the sides of the conduit, this is accomplished with a triangular pressure distribution over only a portion of the bottom slab such that the centroid of the triangular load is colinear with the center of the applied wheel load. If the specified live load condition is the HL = 93 load, then analyses must be completed for both the design truck and the design tandem. Analyses must also be completed for wheel loads at multiple locations across the top slab to determine the maximum moment, thrust, and shear forces at each design location.

Summing load cases – After the analysis is complete, the design forces from the individual load cases are factored and summed into appropriate load combinations to determine the controlling design forces at each location. Structure weight, vertical earth load, and minimum horizontal earth loads are considered permanent dead loads, while the maximum additional lateral pressure (maximum minus minimum), internal water pressure, and live loads are considered temporary and are only considered if they increase the design force at the design location under consideration. Figure 7.12 presents the factored moment envelopes for the maximum and minimum load conditions for a 12 ft by 8 ft (3.6 m by 2.4 m) four-sided box conduit with 10 ft (3 m) of cover and a design truck live load. The sign convention has tension as positive on the inside of the conduit. The figures are only plotted between the faces of the walls and slabs. Positive and negative moments in the slabs are plotted in the regions where they could control the reinforcement design.

Items demonstrated in Figure 7.12, some of which have been previously mentioned, include the following:

- Maximum positive moments in the top and bottom slabs occur with maximum vertical loads and minimum lateral loads. Maximum positive moments in the sidewall occur with maximum lateral loads and minimum vertical loads.
- Peak negative moments generally occur under maximum vertical and horizontal loads, excluding the internal water load.
- Peak negative moments in the top and bottom slabs are relatively small at the tip of the haunch, as this is near the inflection point.
- In this model, the largest negative moments occur in the sidewall, which is typical for these structures.

- There is no positive moment in the sidewall. Most of the moments in the sidewalls are due to the high moments at the conduit corners due to vertical loads.

Most box conduits are square or have spans larger than the rise. If the rise is greater than the span, the side moments increase due to the extra lateral pressure, and there are likely to be positive moments in the sidewall. Multicell box conduits have similar moment diagrams, except the negative moments where the slabs pass over the interior walls can have high negative moments.

Figure 7.13 shows the peak factored moments for a 12 ft by 8 ft (3.6 m by 2.4 m) three-sided box conduit with the same loading conditions as the box in Figure 7.12. The three-sided box has no significant positive moments in the sidewall and no moment at all at the bottom of the sidewall, which is typically modeled as a pinned connection. This results in a reduction of the top slab positive moments and an increase in the negative moments relative to the four-sided box. Since the peak design negative moments occur in the sidewall at the tip of the haunch, the effect on outside reinforcement is small, and there is an overall saving in total reinforcement relative to a four-sided box conduit. Some programs allow horizontal motion of the footing under lateral soil pressures.

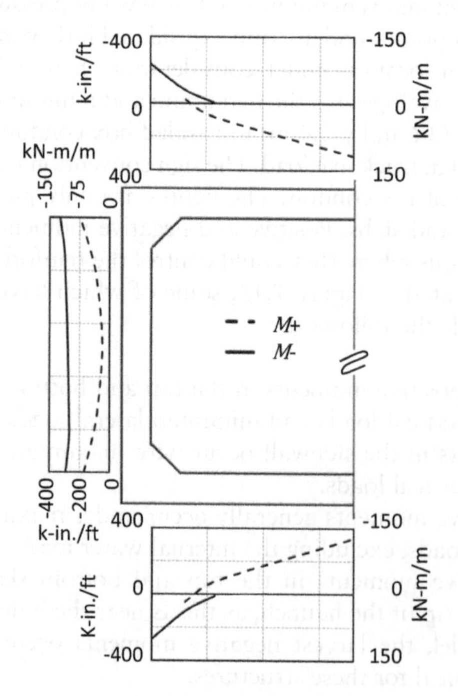

Figure 7.12 Typical moment diagram for 4-sided box conduit.

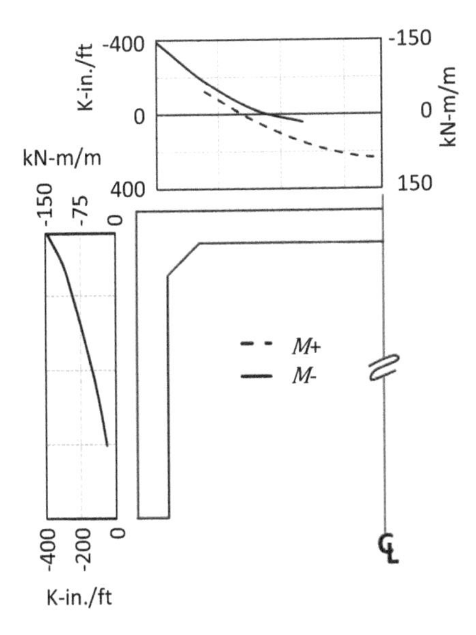

Figure 7.13 Typical moment diagram for three-sided box conduit.

7.5.1.3 Reinforcement design and layout

Reinforcement design for box conduits is completed using standard design equations for cracking, flexure, and shear strength, which are presented in AASHTO LRFD Article 5. While the reinforcing steel in box conduits could be designed by the equations presented in AASHTO LRFD Article 12.10 for round concrete conduits (see Section 7.5.3), these latter equations are not always accepted as being appropriate for box conduits.

There are many possible configurations of reinforcement for box conduits. Figure 7.14 shows a common configuration for precast box conduits with more than 2 ft of fill. These boxes are cast on end, and all reinforcement can be placed prior to adding concrete. Precast box conduits are generally reinforced with welded wire reinforcement; thus, the slab and sidewall reinforcements cannot overlap due to interference of the longitudinal wires. In this figure, the U-shaped cages reinforce the peak negative moment locations while the straight reinforcement addresses the positive moment locations. Some specifiers require reinforcement of all surfaces, which can be achieved by extending the U-shaped reinforcement or providing additional straight reinforcement between the ends of the U-shaped reinforcement. The reinforcement layouts for precast box conduits should always be assessed for adequacy during handling. If the lifting method results in tension on the inside of the corners of the box conduit, additional reinforcement may be required.

Figure 7.15 shows one possible configuration for a cast-in-place box conduit where the bottom slab is commonly placed separately from the walls

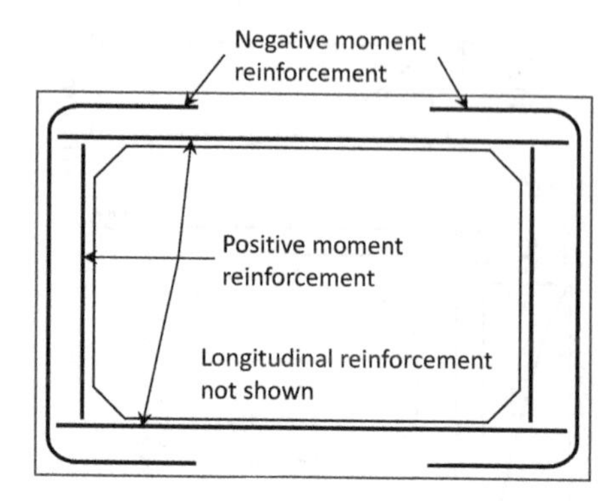

Figure 7.14 Common reinforcement configuration for precast box conduits with depth of fill greater than 8 ft, 2.4 m.

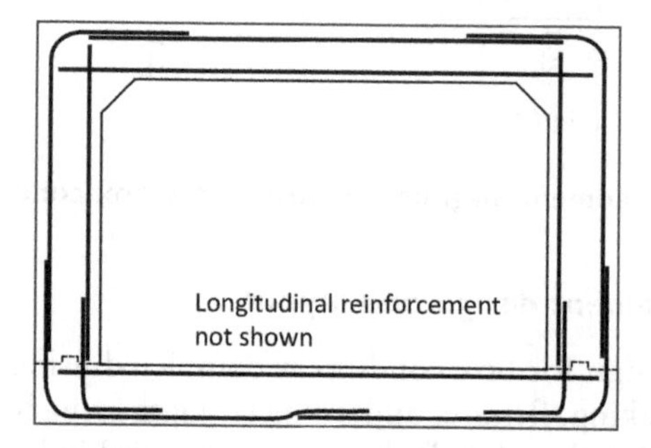

Figure 7.15 Possible reinforcement configuration for cast-in-place box conduits.

and top slab. Cast-in-place box conduits are typically reinforced with deformed bars. No bottom haunches are shown in Figure 7.15, which makes forming more straightforward, but the structural benefits of the haunch are lost.

7.5.2 Circular, elliptical, and arch concrete conduits

Reinforced concrete conduits are produced in circular, elliptical, and arch shapes (Figure 3.1). Elliptical conduits are designed and installed in both horizontal and vertical configurations. There are two primary approaches to designing these conduits. Indirect design is the oldest method, using semi-empirical procedures to determine load and the required strength in a three-edge bearing (TEB) test. The direct design procedure models a buried conduit for field installation conditions and determines the in-ground moment, thrust, and shear forces for which reinforcement requirements can then be determined directly by calculation. For direct design, conduit forces can be calculated using one of the pressure distributions described in Chapter 5 or by using finite element analysis, which allows a designer much greater freedom in modeling installation conditions.

7.5.2.1 *Indirect design*

The first design method developed for concrete conduits was the indirect method. In terms of modern specifications, parts of the original indirect design method are out of date. For example, the descriptions of soils and compaction levels were vague. The soil descriptions cannot be definitively identified in terms of ASTM D2487, ASTM D2321, or other standards, and "well compacted" does not define a unit weight that complies with a specific Proctor density. However, in times past, engineers and contractors were able to reach an agreement on these terms and produce good installations. For modern specifications, the method has been updated to address these issues. Indirect design has been successfully applied for many years and has the benefit of producing a TEB capacity for a conduit that can be used for quality control testing. Since the TEB test is conducted until a crack opens to a width of 0.01 in. (0.3 mm) or ultimate failure, it can offer strengths greater than predicted by calculation. This is in part because testing takes advantage of concrete strengths and reinforcement yield strengths being higher than the required minimum, and curvature effects that increase capacity over straight beam theory. Studying the indirect method offers insights into buried conduit behavior and design that are not always readily apparent with other design approaches.

Indirect design consists of determining the loads, the required conduit strength, and a bedding factor, all of which were briefly introduced in Chapter 2. This section provides more detail on the method.

Indirect design – trench loads: Earth loads are generally calculated using the Marston load theory, which for trenches is illustrated in Figure 7.16. The vertical shear stresses on the trench wall from settlement of the backfill and the horizontal stresses due to soil lateral pressure transfer a portion of the backfill load to the in situ soils in the trench wall. This theory assumes no load is carried by the backfill at the sides of the conduit. This was likely based on practice at the time, which did not require compacting sidefill and that the conduit was so much stiffer than the sidefill that it would support most of the weight of the trench backfill. Figure 2.5 demonstrates that their testing procedures did not allow for any sidefill. The trench load equations are

$$W_E = C_d \gamma_s B_d^{\,2}, \tag{7.4}$$

$$C_d = \frac{1 - e^{-2K_a \mu' \left(\frac{H}{B_d}\right)}}{2K_a \mu'}, \tag{7.5}$$

where

W_E = weight of earth carried by the conduit, lb/ft, kN/m
C_d = load coefficient, dimensionless (see Figure 7.17)
γ_s = unit weight of soil, lb/ft^3, kN/m^3

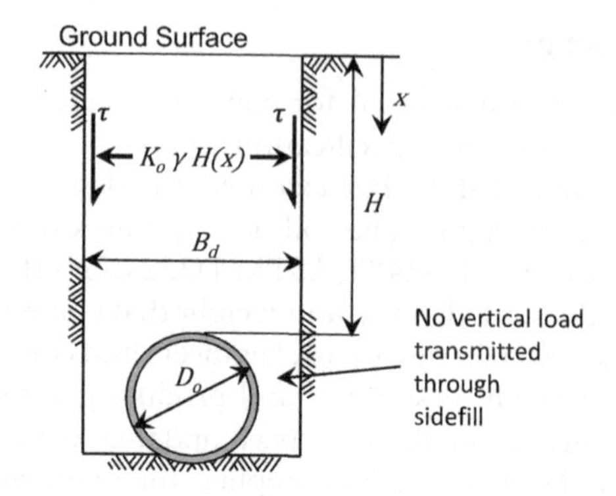

Figure 7.16 Marston trench load assumptions.

B_d = trench width, ft, m
K_a = active lateral pressure coefficient
μ' = tan ϕ', coefficient of friction between the fill material and the trench wall
ϕ' = soil friction angle
H = depth of fill over the conduit, ft, m
e = base of natural logarithms

Figure 7.17 shows values of C_d for five soil types that Marston felt represented soils used in practice. Detailed soil descriptions, such as gradation, fines content, and Atterberg limits, were not used when the method was developed. Figure 7.17 shows that C_d is approximately equal to H/B_d for very shallow trenches which means the entire weight of the backfill soil is carried by the conduit. As the depth of fill increases for a constant trench width, the arching increases due to frictional and horizontal forces at the trench wall/backfill interface; hence, the proportion of the load carried by the conduit decreases, and the lines for each soil curve upward, indicating greater arching. The product $K_o\mu'$ represents the parameters that control the maximum frictional force that can be developed. The load reduction for conduits with high ratios of H/B_d is significant; however, installing conduit in deep, narrow trenches can be challenging from a construction point of view.

It is easier to see the difference between different load conditions by expressing the load on a conduit in terms of the vertical arching factor, called VAF_t for trench loading:

$$W_E = VAF_t \, H \, \gamma_s \, B_c. \tag{7.6}$$

Substituting Equation 7.4 into 7.6 and rearranging yields:

$$VAF_t = C_d \left(\frac{B_d}{H}\right)\left(\frac{B_d}{B_c}\right), \tag{7.7}$$

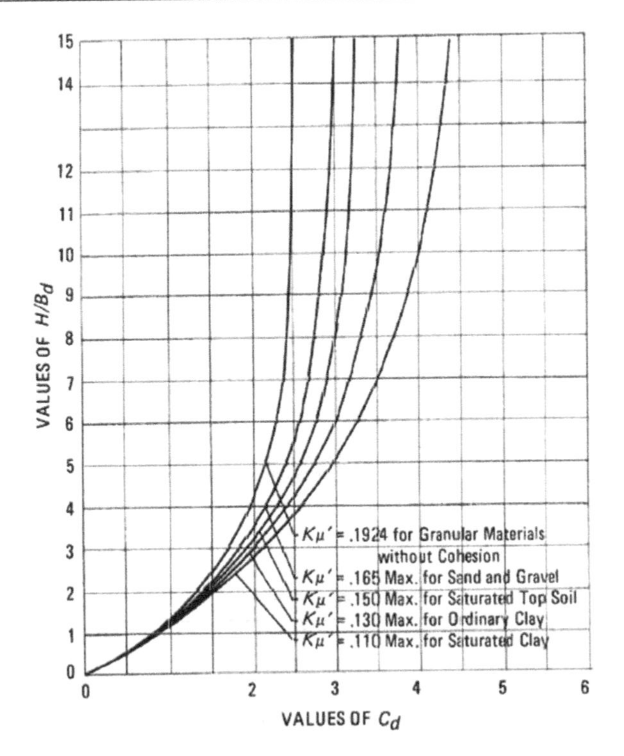

Figure 7.17 Load coefficient for trench installations. (ACPA 1998, with permission)

where

VAF_t = vertical arching factor for the trench condition

The vertical arching factors for a 6 ft (1.8 m) outside diameter conduit embedded in backfill with $K\mu' = 0.165$ in two different width trenches are presented in Table 7.1, which shows the load reduction possible in a narrow trench. When B_d/B_c is 1.5, VAF_t is greater than 1, except for the deepest condition. Reducing B_d/B_c to 1.17 reduces the load on the conduit by about 23%–30%; however, designers and installers must determine if backfill can be worked into the haunch zone and compacted properly in the narrow trench.

Indirect design – embankment loads: Loads on concrete conduits in embankment conditions are calculated in a similar manner, but the load transfer on or off the conduit is assumed to occur across the vertical planes above the springlines. The shear forces in the soil that develop from the settlement of the sidefill can increase or decrease the load. In most embankment load cases, the shear forces increase the load on the conduit, as shown in Figure 7.18.

Table 7.1 VAF_t versus H/B_d for a 6 ft (1.8 m) conduit

	$B_d/B_c = 1.5$ (9 ft, 2.7 m trench width)			$B_d/B_c = 1.17$ (7 ft, 2.1 m trench width)		
H, ft (m)	H/B_d	C_d	VAF_t	H/B_d	C_d	VAF_t
1 (0.3)	0.11	0.11	1.48	0.14	0.14	1.14
2 (0.6)	0.22	0.21	1.45	0.29	0.27	1.11
6 (1.8)	0.67	0.60	1.35	0.86	0.75	1.02
9 (2.7)	1.00	0.85	1.29	1.29	1.05	0.95
18 (5.5)	2.00	1.46	1.10	2.57	1.73	0.79
27 (8.2)	3.00	1.90	0.95	3.86	2.18	0.66

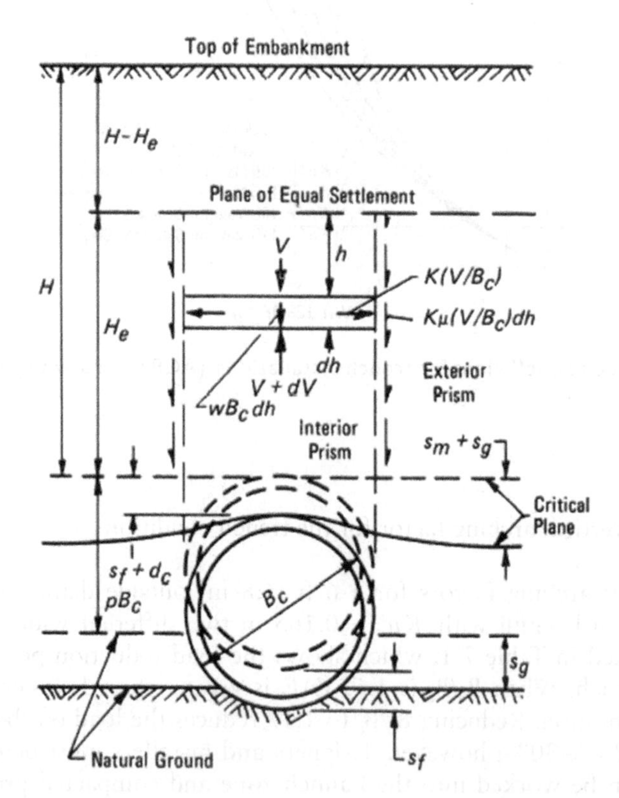

Figure 7.18 Settlements that influence loads on conduits in embankments. (ACPA 1998, with permission)

The key parameter affecting load is the critical plane – the horizontal line passing over the top of the conduit before backfilling. The shape of this plane after backfilling is indicative of the earth load carried by the conduit. As shown in Figure 7.18, the sidefill between the top of the in situ soil and the top of the conduit, called the projection height ($p\ B_c$, where p is the projection ratio), compresses an amount s_m, creating shear forces that increase the embankment vertical arching factor (VAF_e) to values greater than 1. Marston defined the settlement of the sidefill relative to the settlement of the top of the conduit as the settlement ratio, r_{sd}:

$$r_{sd} = \frac{(s_m + s_g) - (s_f + d_c)}{s_m} \tag{7.8}$$

$VAF_e = 1.0$ if the critical plane remains a straight line. The condition shown in Figure 7.18 gives a positive value of r_{sd} and VAF greater than 1.0. If the top of the conduit settles more than the sidefill, r_{sd} is negative, and VAF is less than 1.0.

The plane of equal settlement is the elevation above which there is no differential settlement in the soil and there are no shear forces altering the load on the conduit. Above the plane of equal settlement, the contribution of load to the conduit has $VAF = 1$. The condition shown in Figure 7.18 is called the incomplete trench condition with the shear forces dissipating prior to reaching the ground surface.

Marston developed Equation 7.9 to calculate embankment loads for the complete and incomplete projection conditions:

$$W_c = C_c \gamma_s B_c^2, \tag{7.9}$$

where

W_c = earth load on conduit for the complete projection condition, lb/ft, kN/m
C_c = load coefficient, which is defined differently for the complete and incomplete conditions

The incomplete trench condition, when the plain of equal settlement is below the ground surface, produces a cumbersome equation for C_c that, at the time, would have been difficult to solve. Marston (1930) addressed this difficulty by taking advantage of the fact that the settlement ratio, r_{sd}, and projection, p, ratio were the key parameters controlling conduit embankment loads. Table 7.2 presents r_{sd} for the positive projecting condition for three bedding conditions that Marston determined affect the settlement ratio in addition to the settlement of the soil sidefill. This table suggests that the bedding support under the conduit is the dominant contributor to the settlement ratio. Settlement of the sidefill is addressed by knowing the projection ratio, p.

Marston then developed Figure 7.19, from which designers could directly read the value of C_C based on the ratio of backfill depth to conduit diameter, H/B_c, and the product of settlement ratio and the projection ratio, $r_{sd}p$.

Table 7.2 Settlement ratios for positive projection conditions

Installation and Foundation Condition	*Settlement Ratio, r_{sd}*	
	Usual Range	*Design Value*
Positive Projection	0.0 to +1.0	
Rock or Unyielding Soil	+1.0	+1.0
[a]Ordinary Soil	+0.5 to +0.8	+0.7
Yielding Soil	0.0 to +0.5	+0.3
Zero Projection		0.0

[a] ACPA (1998) states, "With construction methods resulting in proper compaction of bedding and sidefill materials, a settlement ratio design value of +0.5 is recommended."

Source: ACPA 1998, adapted with permission.

This figure shows that shallowly buried conduits with low H/B_c ratios are at or near the complete condition, and deeper conduits will mostly be in the incomplete condition.

As with the trench load solution, it is convenient to interpret results by expressing Equation 7.9 in terms of the soil prism load, VAF_e, which is a simple expression:

$$VAF_e = \frac{C_c}{H / B_C} \tag{7.10}$$

A linear relationship between C_c and H/B_c indicates a constant value of VAF_e for the straight-line portions of Figure 7.19, as presented in Table 7.3.

The value of $r_{sd}p = 0.3$ yields VAF_e 1.4. As seen next, the newer direct design method concluded that $VAF_e = 1.4$ for most installations where backfill is compacted, and the bedding is not hard.

Solutions for the negative projecting and induced trench installations and for elliptical conduits are not presented here, as they are not in wide use. ACPA (1998) presents this information.

Indirect design – supporting strength: Once the loads on a conduit are calculated, the next step in indirect design is to determine the required conduit capacity in a TEB test (Section 3.2.2). The basic equation for this is

$$TEB = \frac{(W_E + W_L)L_f}{B_f}, \tag{7.11}$$

where

TEB = Required conduit capacity in a TEB test, lb/ft, N/m
W_E = Earth load on the conduit, lb/ft, N/m
W_L = Live load on the conduit, lb/ft, N/m
L_f = Load factor
B_f = Bedding factor

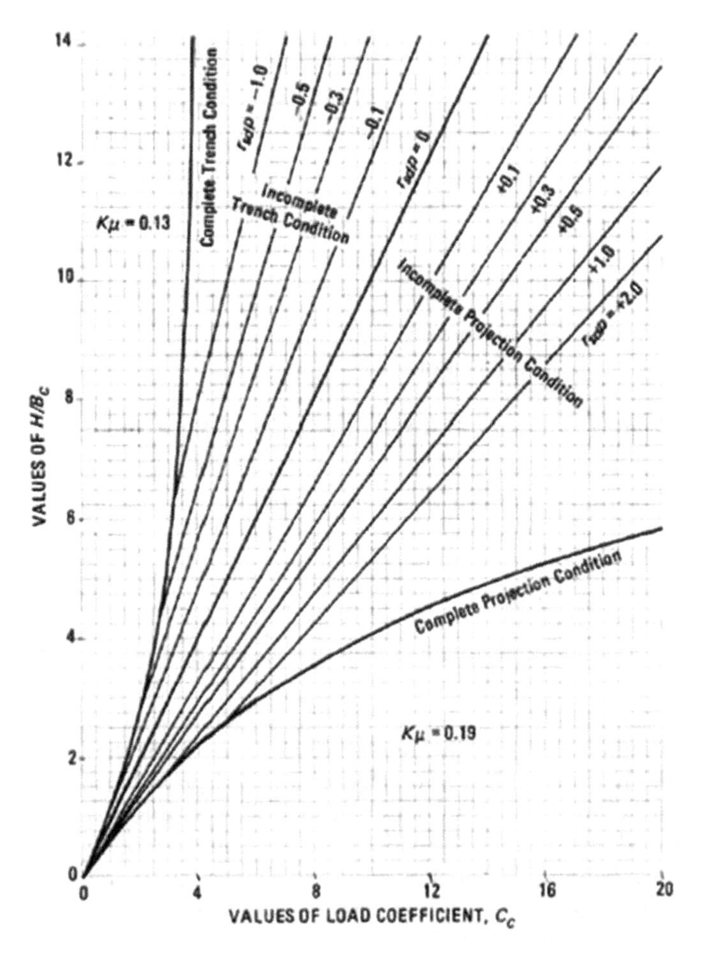

Figure 7.19 Load coefficient, C_c, for positive projecting conditions. (ACPA 1998, with permission)

Table 7.3 $r_{sd}p$ versus VAF_e for incomplete projection conditions

$R_{sd}p$	VAF_e
0.1	1.2
0.3	1.4
0.5	1.5
1.0	1.7
2.0	1.9

As introduced in Section 2.2.4, the bedding factor is the ratio of the in-ground moment to the moment in a TEB test under the same load. It is further convenient to normalize the required TEB strength by dividing TEB by the conduit diameter. This is called the D-load:

$$DL = TEB / D_i, \tag{7.12}$$

where

DL = required D-load, lb/ft/ft, N/m/mm
D_i = conduit diameter, ft, mm

The required D-load at the service and factored conditions include the following:

$DL_{0.01}$ = D-load at the occurrence of a 0.01 in. crack in a TEB test, the common service load limit in US units
$DL_{0.3}$ = D-load at the occurrence of a 0.3 mm crack in a TEB test, the common service load limit in SI units
DL_u = D-load required for the strength load condition

The use of D-loads allows for grouping concrete conduits into classes where a common D-load indicates sufficient strength for the same installation conditions independent of diameter. Table 7.4 provides the D-loads for the standard classes, which are typically identified with Roman numerals.

Trench bedding factors – Bedding factors were introduced in Section 2.2.4. Trench bedding factors are related to the installation conditions and, because of the low lateral pressures assumed in the trench load model, are constant, as shown in Figure 7.20.

Table 7.4 Standardized D-load classes for concrete conduits

	US Units (lb/ft/ft)		*SI Units (N/m/mm)*	
Class	$DL_{0.01}$	DL_u	$DL_{0.3}$	DL_u
I	800	1,200	40	60
II	1,000	1,500	50	75
III	1,350	2,000	65	100
IV	2,000	3,000	100	150
V	3,000	3,750	140	175

Note: Long-standing practice uses L_f = 1.0 for $DL_{0.01}$ and $DL_{0.3}$ for all classes, and L_f = 1.5 for DL_u in Class I – IV conduits and L_f = 1.25 for DL_u Class V conduits

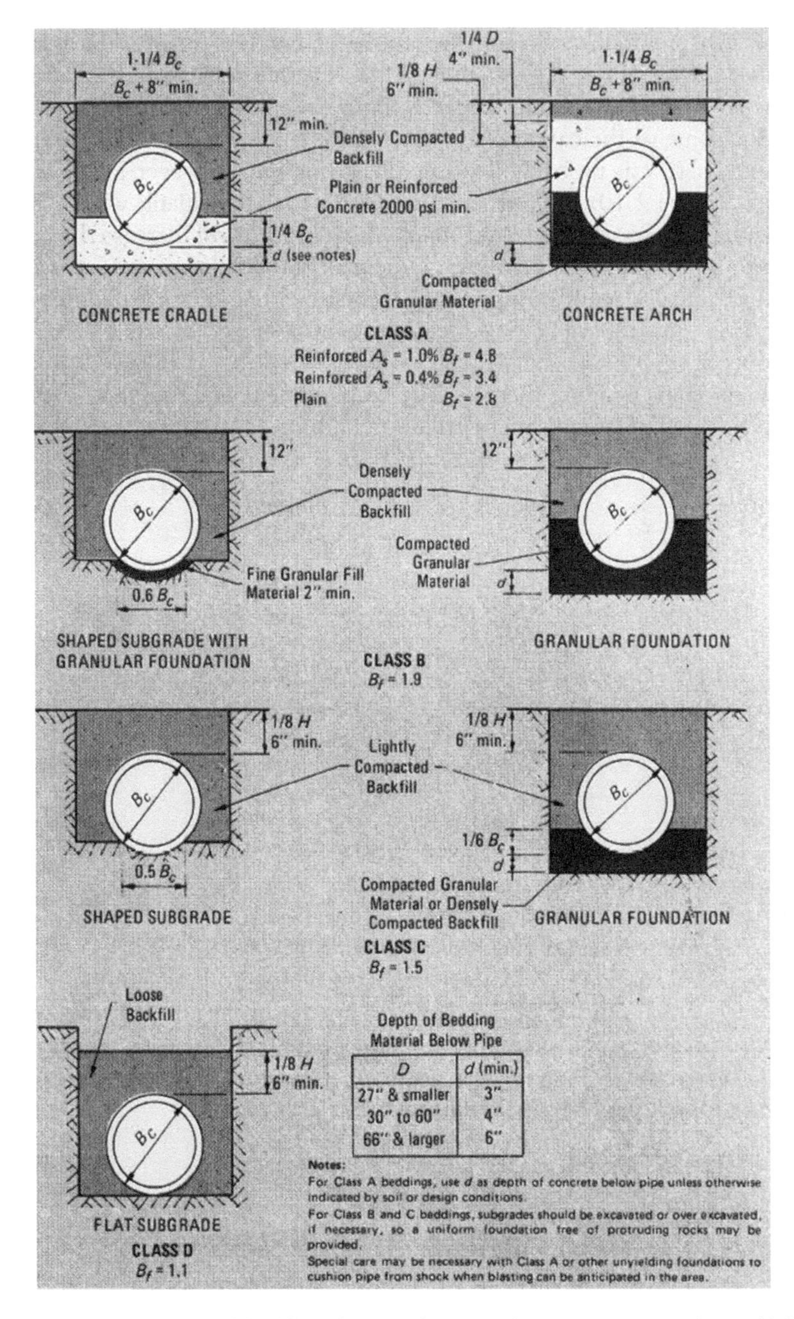

D	d (min.)
27" & smaller	3"
30" to 60"	4"
66" & larger	6"

Figure 7.20 Trench load bedding factors for circular concrete conduits. (ACPA 1998, with permission)

One difficulty in the use of a constant trench bedding factor is the assumption that the weight of the trench backfill is carried solely by the conduit or the trench wall. This assumption was likely based on the test configuration used for some of the early tests (Figure 2.5), where the trenches were narrow, and there was no backfill below the springline. However, as the trench widens (Figure 2.10), it is increasingly likely that the sidefill will carry an increasing portion of the vertical load, which in turn will increase the lateral pressure and reduce the bending moment in the conduit, i.e., increase the bedding factor. A modification of the trench bedding factor to address this concern is introduced next after the embankment bedding factor is introduced.

Embankment bedding factor – The embankment bedding factor for the positive projection condition of circular conduits is calculated as

$$B_{fe} = \frac{A}{N - xq} \tag{7.13}$$

with

$$q = \frac{mK_o}{C_c}\left(\frac{H}{B_c} + \frac{m}{2}\right), \tag{7.14}$$

where

B_{fe} = embankment bedding factor
A = constant, 1.43 for circular conduits
N = constant based on the bedding assumption, Table 7.5
x = parameter based on m, Table 7.6
q = ratio of total lateral load to total vertical load on the conduit
m = fractional portion of the outside diameter of the conduit over which lateral pressure is effective; usually taken equal to the projection ratio, p
K_o = coefficient of lateral earth pressure at rest; often assumed to be 0.33 for bedding factor calculations

Table 7.5 Parameter N for the bedding factor

Bedding Class	N
B	0.707
C	0.840
D	1.310

Source: ACPA 1998, adapted with permission.

Table 7.6 Parameter *x* for the bedding factor

m	*x*
0.0	0.000
0.3	0.217
0.5	0.423
0.7	0.594
0.9	0.655
1.0	0.638

Source: ACPA 1998, adapted with permission.

ACPA (1998) has additional information for calculating bedding factors for other conditions and for noncircular conduits. Note that horizontal and vertical elliptical conduits have significantly different bedding factors because of the difference in the ratio of vertical to horizontal span.

Trench bedding factor modification – A later modification to Marston's theory was to allow the bedding factor to transition from the trench bedding factor to the embankment bedding factor as the vertical load transitions from the trench load to the embankment load (ACPA 1998):

$$B_{fv} = \left(B_{fe} - B_{ft}\right)\left[\frac{B_d - \left(B_c + k_u\right)}{B_{dt} - \left(B_c + k_u\right)}\right] + B_{ft}, \tag{7.15}$$

where

B_{fv} = variable trench bedding factor for design
B_{fe} = embankment bedding factor
B_{ft} = trench bedding factor
B_c = Outside conduit diameter, ft, m
B_d = trench width, ft, m
B_{dt} = trench transition width, ft, m
k_u = factor for minimum trench width, 1.0 ft, 0.3 m

Equation 7.15 provides a bedding factor, B_{fv}, that linearly transitions from the trench bedding factor, B_{ft}, at a trench width of B_c + 1 ft, 0.3 m, to the embankment bedding factor, B_{fe}, at the transition trench width, B_{dt}. Equation 7.15 was not developed rigorously, but it provides a reasonable approach to avoid a stepwise change in the bedding factor at the transition width. Computer programs, such as PIPEPAC (ACPA/CCPA 2023), complete these calculations.

In addition to the trench and embankment bedding factors, Equation 7.15 requires the transition width. The equations for this are cumbersome and require an iterative process to solve. In lieu of a rigorous solution,

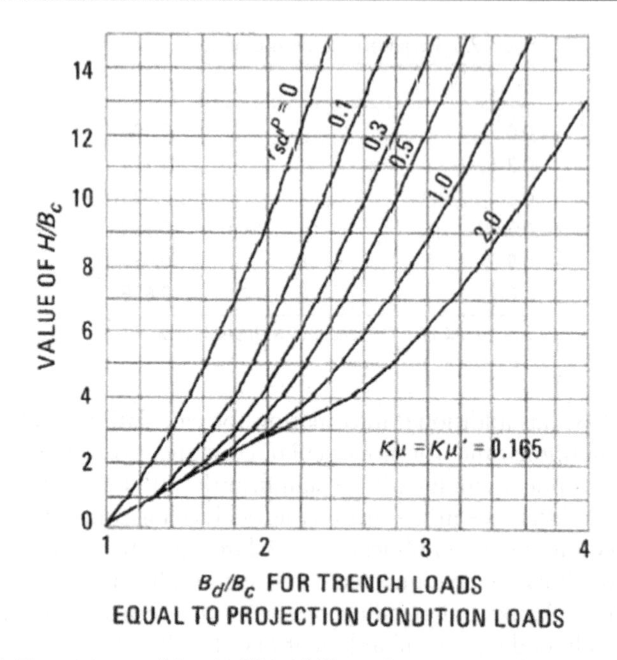

Figure 7.21 Transition widths. (ACPA 1998, with permission)

ACPA (1998) developed Figure 7.21, which provides an approximation of B_d/B_c at the transition width. The figure is based on only one soil type but in practice is used for all soils.

Standard installation bedding factors – Analysis with the Heger Pressure distribution was used to develop "standard installations" for embankment (Figure 7.22, Table 7.7) and trench installation conditions. The four standard installations introduced in Section 5.2.2 use soil descriptions and compaction requirements in terms of modern specifications. While originally developed for direct design, discussed below, standard installation bedding factors have been developed for the indirect design method (Table 7.8).

Trench bedding factors have also been developed for the standard installations (Table 7.9). Similar to the approach with the traditional bedding factors, the trench bedding factors are modified depending on trench width, but the equation provided is different from Equation 7.15. The reason for this difference is not clear. A subscript "s" is added to distinguish the two computations of B_{fv}.

$$B_{fv-s} = \frac{\left(B_{fe} - B_{fo}\right)\left(B_d - B_c\right)}{\left(B_{dt} - B_c\right)} + B_{fo} \tag{7.16}$$

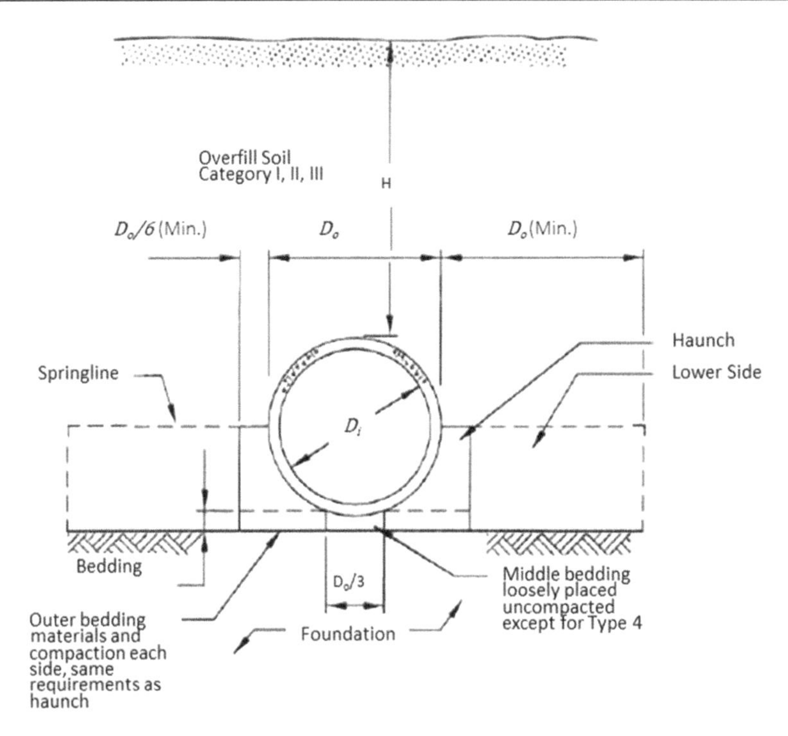

Figure 7.22 Standard embankment installations. (ACPA 1998, with permission)

Table 7.7 Standard embankment installations soil and minimum compaction requirements

<table>
<tr><th>Installation Type</th><th>Bedding Thickness</th><th>Haunch and Outer Bedding[a]</th><th>Lower Side[b]</th></tr>
<tr><td>Type 1</td><td rowspan="4">$D_0/24$ minimum, not less than 3 in. (75 mm).
If rock foundation, use $D_o/12$ minimum, not less than 6 in. (150 mm)</td><td>95% Category I</td><td>90% Category I,
95% Category II,
or
100% Category III</td></tr>
<tr><td>Type 2</td><td>90% Category I,
or
95% Category II</td><td>85% Category I,
90% Category II,
or
95% Category III</td></tr>
<tr><td>Type 3</td><td>85% Category I,
90% Category II,
or
95% Category III</td><td>85% Category I,
90% Category II,
or
95% Category III</td></tr>
<tr><td>Type 4</td><td>No compaction required for Category I and II,
85% Category III,</td><td>No compaction required for Category I and II,
85% Category III,</td></tr>
</table>

[a] Percentage is the required percentage of maximum standard Proctor density for the backfill.
[b] Category references the soil type (see Table 4.9).

Source: ACPA 1998, with permission.

Table 7.8 Bedding factors, B_{fe}, for standard installation embankment conditions

	Standard Installation			
Conduit Diameter	*Type 1*	*Type 2*	*Type 3*	*Type 4*
12 in. (300 mm)	4.4	3.2	2.5	1.7
24 in. (600 mm)	4.2	3.0	2.4	1.7
36 in. (900 mm)	4.0	2.9	2.3	1.7
72 in. (1,800 mm)	3.8	2.8	2.2	1.7
144 in. (3,600 mm)	3.6	2.8	2.2	1.7

Source: ACPA 1998, with permission.

Table 7.9 Bedding factors, trench conditions, B_{fo}

Standard Installation	*Minimum Bedding Factor*
Type 1	2.3
Type 2	1.9
Type 3	1.7
Type 4	1.5

Source: ACPA 1998, with permission.

7.5.2.2 Direct design

Direct design evaluates the forces in a concrete conduit due to an assumed pressure distribution or through finite element analysis. Reinforcement is designed directly for these forces. Finite element modeling and the pressure distributions discussed in Section 5.2.2 are all used for the direct design of concrete conduits. Direct design avoids the need to provide reinforcement necessary to pass the three-edge bearing test that is not needed to resist in-ground forces; however, this is not always desirable, as the TEB test is the preferred method of post-production quality control.

7.5.2.3 Indirect versus direct design

Much of the value of indirect design lies in the importance of the TEB test for quality control purposes. Indirect design determines the loads and bedding conditions of a concrete conduit installation and calculates a required TEB strength that will provide good service in those conditions based on bedding factors. Reinforcement design for the TEB test is based on the forces at the crown and invert of the conduit. The outside reinforcement area is set at 60% of the inside reinforcement area, which is more than required for the TEB test and meets the requirements of all field conditions. Direct design addresses the in-ground load conditions, and the conduit can be designed for expected moment, thrust, and shear at all critical locations.

The difference between the field and TEB condition is demonstrated by comparing the moment, thrust, and shear coefficients around a conduit for the TEB test (Equations 3.12, 3.13, and 3.14, which assume a uniform moment of inertia, i.e., uncracked section properties) with those for a conduit with Olander 90° bedding, which is often compared to traditional B bedding. Olander and any other direct design method forces can be stated in terms equivalent to the TEB equations:

Moment:

$$M = c_m WR \tag{7.17}$$

Shear:

$$V = c_v W \tag{7.18}$$

Thrust:

$$N = c_n W \tag{7.19}$$

where W is the applied load, and R is the conduit radius. The comparison is presented in Figure 7.23

The indirect design method calculates the bedding factor as the ratio of $c_{m\text{-}TEB}$ to $c_{m\text{-}Olander}$. For the condition shown, $B_f \sim 0.28/0.14 = 2.0$, which is similar to the Marston bedding factor for Type B bedding. However, the ratio of TEB to Olander coefficients at all other locations is not two. In particular:

- While the moments at the invert are close to the in-ground condition with $B_f = 2$, the thrust forces, which reduce the required reinforcement, are zero in the TEB test.
- Shear forces in the TEB test are highest at the support points where in-ground shear forces are zero. The Olander peak shear force is located away from the invert and has a lower magnitude. The result of this is that some conduits may require stirrup reinforcement for the TEB test but not for the in-ground condition.
- At the springline (90° from invert), the ratio of the peak negative moments is approximately 2, matching the bedding factor, but the ratio of the thrusts is closer to 1.0; thus, the TEB test does not test the negative moment capacity of the conduit prior to yielding the invert and crown reinforcement.

Despite the differences between the in-ground condition and the TEB test, indirect design remains a valuable tool for routine designs, and the ability to

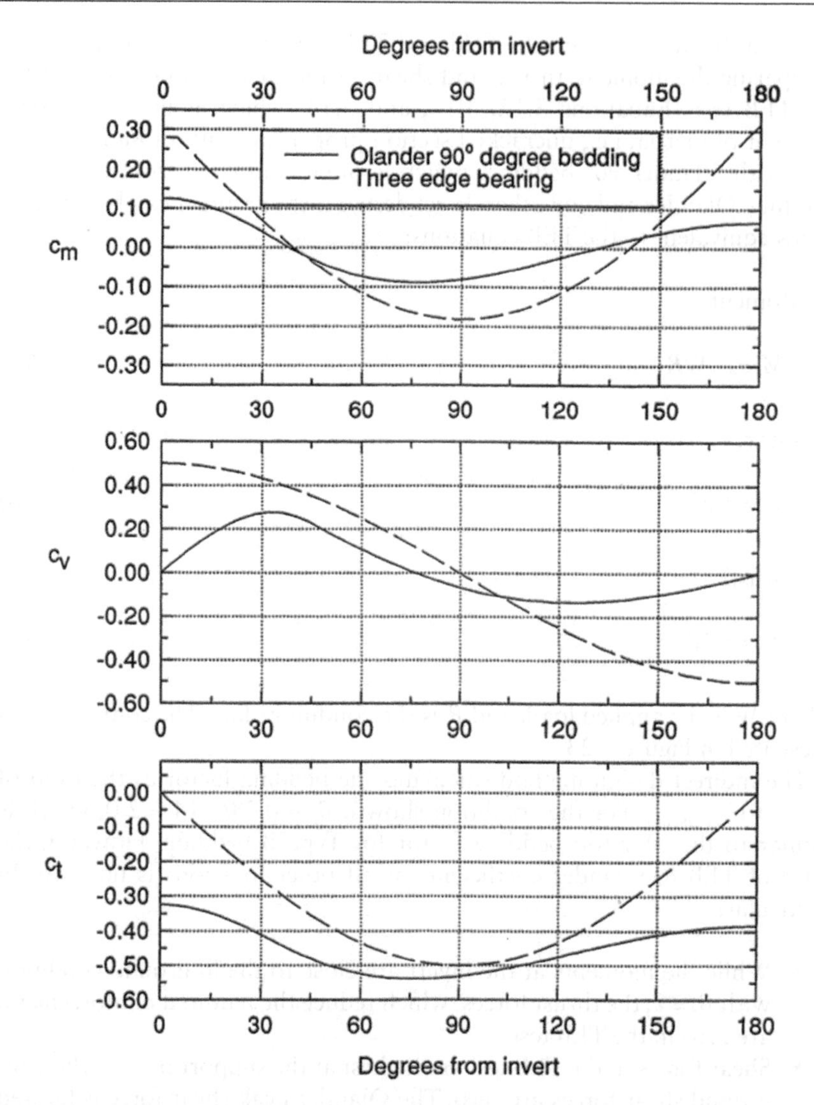

Figure 7.23 Comparison of force coefficients for the TEB test and Olander 90° bedding prior to cracking.

conduct TEB quality control tests is important as it tests the fully manufactured conduit. In the absence of TEB testing, quality control would depend on monitoring the manufacture and/or coring of the conduit to verify concrete strength and reinforcement quantity and placement.

Indirect design is particularly important for small-diameter conduits. The direct design equations for reinforcement design are based on traditional beam theory, which is adequate for designs with inside diameters of 48 in. (1,200 mm) and larger but become increasingly conservative for smaller diameters. AASHTO LRFD includes a moment modifier (Moore et al., 2014) that corrects the moment at the crown for some of the conservatism. This correction addresses analysis errors for calculations based on thin rings, while concrete conduits are actually thick rings.

$$M_{thick} = M_{thin}\left(1 - 0.373t/R\right) \tag{7.20}$$

Moore et al. (2014) demonstrated that the effect of the burial conditions on the correction is modest, and it is appropriate for crown moment obtained from any analysis based on thin ring theory (finite element analyses using structural elements based on "thin beam" theory).

7.5.3 Reinforcement design of concrete conduits

Reinforcement design for concrete conduits can be completed with standard methods used for general reinforcement design; however, much research has been conducted on reinforcement design for round concrete conduits to address specific combinations of forces and the effects of curvature. The equations presented in AASHTO LRFD Article 12.10 for these conduits appear non-standard but are applicable to all reinforced concrete sections. The reinforcing design equations for nonrectangular conduits apply to direct and indirect design, meaning they are applicable to both in-ground conditions and TEB test conditions. Development of the concrete conduit design equations in AASHTO LRFD is documented in Heger and McGrath (1982, 1982a, 1983, 1984).

7.5.3.1 Design for service limit state – crack width control

Controlling crack width is a service limit condition to minimize environmental exposure and thus corrosion of the reinforcement. Crack width is affected by the reinforcement spacing and the depth of clear concrete cover because cracks widen with increasing distance from points of restraint. The type of reinforcement used can also affect the width of crack. AASHTO LRFD treats crack control of rectangular and curved concrete conduits differently.

Box Conduits - AASHTO LRFD Article 5.6.7 provides equations 7.21 and 7.22 (AASHTO LRFD Eqs. 5.6.7-1 and 5.6.7-2):

$$s \le \frac{700\gamma_e k_u}{\beta_s f_{ss}} - 2d_c, \tag{7.21}$$

$$\beta_s = 1 + \frac{d_c}{0.7\left(h - d_c\right)}, \tag{7.22}$$

where

s = reinforcement spacing, in., mm
γ_e = exposure factor, 0.75 for buried conduits
β_s = ratio of flexural strain at the tension face to the strain at the centroid of the reinforcement layer nearest the tension face
f_{ss} = calculated service load tensile stress in the reinforcement, ksi, MPa $\leq 0.6\ F_y$
F_y = reinforcement yield stress, ksi, MPa
k_u = unit correction, 1.0 for US units, 175 for SI units
d_c = concrete cover measured from the tension face to the centroid of the closest reinforcement layer, in., mm
h = overall thickness of conduit wall or slab, in. mm

The AASHTO LRFD Commentary accompanying the equations suggests that with $\gamma_e = 0.75$, the limiting crack width is about 0.013 in. (0.32 mm). The traditional limit for crack width in concrete conduits is 0.01 in. (standards in SI units use 0.3 mm). The Commentary also states, "There appears to be little or no correlation between crack width and corrosion." This statement is true, but there is consensus that crack width should be controlled, and the limit of approximately 0.01 in. (0.3 mm) remains generally accepted for concrete conduits.

The basis for the requirement to limit the service load reinforcement stress to 0.6 F_y is unclear. This limit is explained in AASHTO LRFD C5.6.7 as being based on a recommendation that the service stress limit for reinforcement with F_y = 100 ksi (690 MPa) should be limited to 60 ksi (410 MPa), which led to the 0.6 F_y limit being applied to all reinforcements. Since crack width is based on strain in the reinforcement and not stress, the concern should be to control strains to less than what occurs at a stress of 60 ksi (410 MPa). Thus, lower yield reinforcements should not require a service stress limit less than 60 ksi (410 MPa) unless required by Equation 7.21.

The aforementioned is supported for conduits by prior research. Precast box conduits, like round concrete sections, typically have low clear depths of cover over reinforcement, commonly 1.0 in. (25 mm), and closely spaced reinforcement (2 in., 50 mm is common for welded wire reinforcement). Heger and McGrath (1982, 1984) studied crack control in round and rectangular conduits during the development of a design equation that is now incorporated into AASHTO LRFD for reinforced concrete conduit design (Article 12.10.4.2.4d). This research showed that the reinforcement stress at the occurrence of 0.01 in. (0.25 mm) crack is often much higher than 0.6 F_y, particularly at low reinforcement ratios, which is also common for pipe and box conduits (Figure 7.24). The figure also demonstrates that reinforcement type affects the reinforcement stress at the occurrence of a 0.01 in. (0.25 mm)

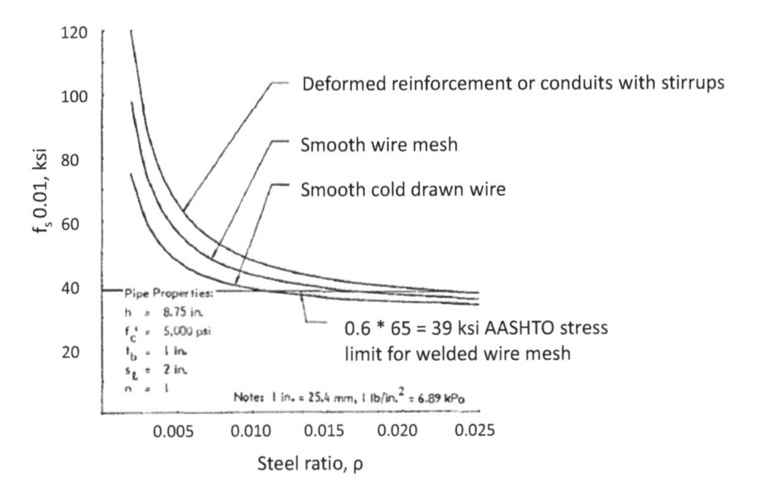

Figure 7.24 Reinforcement stress at 0.01 in. (0.25 mm) crack versus reinforcement ratio (Adapted from Heger and McGrath 1982; and Heger and McGrath 1984. Authorized reprint from *Proceedings ACI Journal*, Vol. 81, No. 2)

crack for common conduit reinforcements. Without the limit of 0.6 F_y box, conduits under deep fills with no live load could have a service stress limit of 65 ksi (450 MPa) /1.3 = 50 ksi (345 kPa).

Nonrectangular concrete conduits – Crack control in nonrectangular concrete conduits (AASHTO LRFD Article 12.10.4.2.4d) was developed from Figure 7.24 and considers the same parameters as discussed for box conduits but also considers the reinforcement type. The form of the equation, slightly rearranged from AASHTO, is

$$A_{sc} = \frac{B_1}{k_1\,\phi\,d\,F_{cr}}\left[\frac{M_s + N_s\left(d - \dfrac{h}{2}\right)}{i\,j} - k_2\,C_1\,b\,h^2\sqrt{f_c'}\right], \tag{7.23}$$

where

A_{sc} = required reinforcement for crack control, in.^{2}/ft, m^2/m
B_1 = coefficient based on reinforcement placement, in., m (units are inches or m even though the calculation suggests in.$^{2/3}$, m$^{2/3}$)

$$= \left(\frac{t_b\,s_l}{2n}\right)^{\frac{1}{3}}$$

t_b = clear cover over reinforcement, in., m
s_l = spacing of reinforcement, in., m
n = number of layers of reinforcement
ϕ = resistance factor
d = depth from compression surface to centroid of tensile reinforcement, in. m
F_{cr} = coefficient to adjust conservatism of the calculation
M_s = service moment, in.-k/ft. MN-m/m
N_s = service thrust force (+ compression), k/ft, MN/m (AASHTO LRFD presents an alternate equation for thrust in tension)
h = conduit wall thickness, in., m

$$j = 0.74 + 0.1\frac{e}{d} - \frac{h}{2} \leq 0.9$$

$$I = \frac{1}{1 - \frac{jd}{e}}$$

$$e = \frac{M_s}{N_s} + d - h/2$$

C_1 = coefficient based on reinforcement type
b = length of design section, 12 in./ft, 1 m/m (the units keep the units balanced for this equation).
k_1 = unit correction 30 k/in for US units, 5.25 MN/m for SI units
k_2 = unit correction 0.0316 for US units, 0.083 for SI units, dimensionless
f_c' = design compressive strength of concrete, ksi, MPa; $\sqrt{f_c'}$ carries units psi, MPa

Equation 7.23 is essentially the service moment modified to reflect the reduction in tension due to compressive thrust, minus the cracking moment, and adjusted for reinforcement type and bond. The coefficient C_1 is determined based on the reinforcement type, having higher values for reinforcements that bond more effectively. Thus, C_1 equals 1.9 for deformed bars or wire or any reinforcement with stirrups, 1.5 for smooth welded wire reinforcement with transverse wires spaced less than 8 in. (200 mm), and 1.0 for smooth wire with cross wires spaced 8 in. (200 mm) or greater. As bond increases, flexural cracks are spaced more closely and do not open as wide. F_{cr} =1.0 results in reinforcement expected to produce a 0.01 in. (0.25 mm) crack at the service condition. Lower values are more conservative, and greater values are less so. If the required area calculated with Equation 7.23 is negative, the section will not crack.

7.5.3.2 Design for strength limit states

Box conduit reinforcement is sized at the locations of peak moment. For almost all designs, peak negative moments occur at the tip of the haunch or, in the case of boxes without haunches or with long flat haunches, at the face of the wall. Peak positive moments (tension on the inside face) occur at or near midspan or midrise. Shear strength is generally controlled near supports where shear forces are greatest (Figure 7.25).

Design locations for round, elliptical, and arch concrete conduits, are shown in Figure 7.26. Inside reinforcement is sized for peak positive moment at the crown and invert and for peak negative moment near the springlines. The exact location of peak negative moment varies with the bedding condition. The controlling condition for shear strength is near the crown and invert and varies with the bedding condition. The controlling location for shear design is not necessarily the location of maximum shear force, as the shear strength is a function of both shear force and reinforcement flexural stress. Radial tension strength is evaluated at locations of peak positive moment.

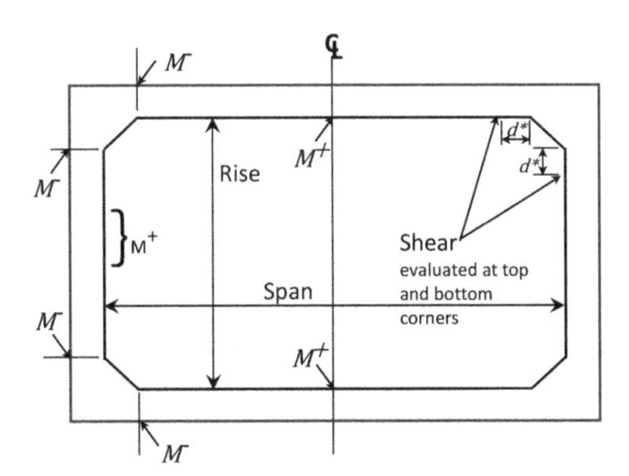

M⁻
℄
Rise
M⁺
d*
M⁻
M⁺
Shear
evaluated at top
and bottom
corners
Span
M⁻
M⁺
M⁻
Note: d*indicates the approximate location of critical shear. The actual dimension varies depending on the load condition and the design code used. "d" is the depth from the tension face of a member to the centroid of the tensile reinforcement.

Figure 7.25 Design sections for moment and shear in box conduits.

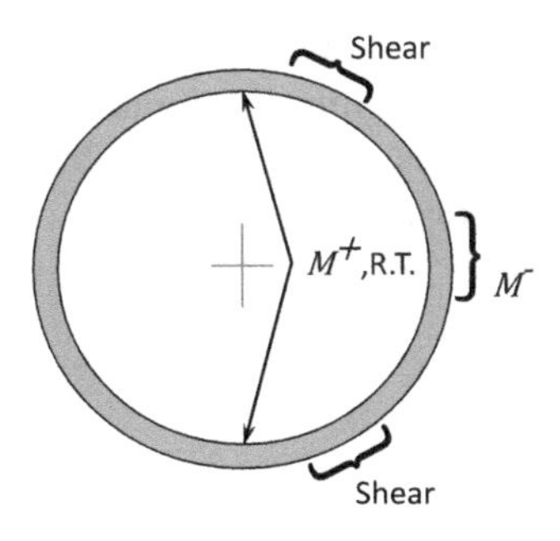

Shear
M⁺,R.T.
M⁻
Shear

Figure 7.26 Design sections for moment (M^+, M^-), shear, and radial tension (RT in concrete conduit).

Once backfilled, nonrectangular concrete conduits and most box conduits do not develop tension on the inside surface of the sidewall, as the compressive thrust stress is greater than the tension due to bending; however, reinforcement may still be required at these locations to withstand handling forces during transportation and installation.

Design for flexure – Reinforcement design for flexure is a common calculation; however, many designs do not consider the benefits of thrust in reducing the tension force in the reinforcement, which can be significant. The design method for flexure (AASHTO LRFD Article 12.10.4.2.4) includes the thrust effect. This equation is also applicable to box conduits:

$$A_s = \frac{g\phi d - N_u - \sqrt{g\left[g(\phi d)^2 - N_u(2\phi d - h) - 2M_u\right]}}{F_y}, \tag{7.24}$$

where

A_s = required reinforcing area for flexure, in²/ft, m²/m
$g = 0.85\, b\, f_c'$, k/(ft-in.), kN/m²
b = length of conduit considered in design, 12 in./ft, 1 m/m (this notation keeps the equation dimensionally consistent)
f_c' = concrete compressive strength, ksi, kPa
d = depth from compression face to centroid of tension reinforcement, in., m
ϕ = resistance factor for flexure
h = depth of member, in. m
N_u = factored thrust at section being designed (+compression), k/ft, kN/m
M_u = factored moment at section being design, k-in./ft, kN-m/m
F_y = yield strength of reinforcement, k/in², kPa

The form of Equation 7.24 is useful for computer calculations. For hand calculations, designers should note the radicand under the radical sign consists of subtracting two large numbers with a small difference. Thus, the calculation must include more significant digits than required for most calculations. For hand calculations, it is often simpler to use the simultaneous equations that are solved iteratively:

$$A_s = \frac{M_u - N_u\left(\dfrac{h-a}{2}\right)}{f_y(d - a/2)} \tag{7.25}$$

$$a = \frac{f_y A_s + N_u}{0.85 b f_c'} \tag{7.26}$$

Maximum flexural reinforcing limits – Two criteria are checked to limit the maximum reinforcement levels for flexure:

- Radial tension stresses (see Section 3.1.1, Figure 3.7) develop on the inside surface of curved concrete members to hold the curved reinforcement in place (Heger and McGrath, 1982, 1983). This stress is resisted by the tensile strength of the concrete and limits the maximum amount of reinforcement that can be developed to resist bending. The design equation adopted by AASHTO LRFD for this condition (12.10.4.2.4c-1) is an approximation, which is dimensionally incorrect, as it assumes b, the standard design width in US practice, is 12 in. Equation 7.27 restores b to the equation:

$$A_{s\,max} = \frac{k_u\, b\, r_s\, F_{rp}\sqrt{f_c'}\,(R_\phi)\,F_{rt}}{f_y}, \tag{7.27}$$

where

$A_{s\,max}$ = maximum reinforcement that can be developed without causing radial tension failure., in.^{2}/ft, m^2/m
b = width of section being designed: 12 in./ft, 1 m/m
k_u = Unit correction 0.0421 for US, 0.111 for SI
r_s = mean radius of curved tension reinforcement, in., m
F_{rp} = a design factor that may be increased above 1.0 if validated by test and approved by the engineer
f_c' = design strength of concrete, ksi, MPa; $\sqrt{f_c'}$ carries units ksi, MPa
R_ϕ = (ϕ_r / ϕ_f) ratio of resistance factors for radial tension and flexure
F_{rt} = a design factor calculated as follows:
For 12 in., 300 mm $\leq S_i \leq$ 72 in., 1,800 mm

$$F_{rt} = 1 + 0.00833(72 - S_i k_1)$$

For 72 in., 1,800 mm $\leq S_i \leq$ 144 in., 3,600 mm

$$F_{rt} = \frac{(144 - S_i k_1)^2}{26{,}000} + 0.8$$

For $S_i \geq$ 144 in. (3,600 mm)

$$F_{rt} = 0.80$$

S_i = inside span of conduit, in., mm
k_1 = 1 for US units, 25.4 for SI units

A fundamental rule of reinforced concrete design is that flexural failures should be ductile, meaning the reinforcement should yield prior to crushing of the concrete. This approach to design assures that cracks can be observed, and steps taken to prevent collapse. If crushing of the concrete occurs first, failure could occur without warning. The maximum amount of reinforcing is thus limited to three-fourths of the amount that would allow crushing.

$$A_{s\,max} = \frac{\dfrac{k_{c1}\, g'\phi\, d}{k_{c2} + f_y} - 0.75\, N_u}{f_y}, \tag{7.28}$$

where

$A_{s\,max}$ = maximum reinforcement that can be developed for flexure, in.^{2}/ft. m^2/m
k_{c1} = constant carrying units: 55 ksi, 380 MPa
k_{c2} = constant carrying units: 87 ksi, 600 MPa
$g' = b\, f_c'\, [0.85 - 0.05\, (f_c'/k_{c3} - 4)]$
$0.85\, b\, f_c' \geq g' \geq 0.65\, b\, f_c'$
b = width of section being designed: 12 in./ft, 1 m/m
k_{c3} = constant carrying units: 1 ksi, 6.9 MPa

Equation 7.28 reduces the maximum allowable reinforcement with increasing thrust. The overall design is still controlled by flexure; thus, ties to restrain the compression reinforcement, which is a requirement of column design, are not required.

Design for Shear Resistance – The current AASHTO LRFD specifications use different methods for evaluating the shear capacity of box and pipe conduits. The methods appear quite different but address shear strength in the same way. AASHTO computes the shear strength of concrete, V_c, and transverse reinforcement (commonly called stirrup or shear reinforcement), V_n, the sum of which are called the nominal shear resistance, V_n. Finally, the factored shear resistance, V_r, is calculated by incorporating the resistance factor; thus,

$$V_r = \phi\, V_n = \phi_v V_c + \phi_v V_{s0}. \tag{7.29}$$

The shear strength from prestressed reinforcement, V_p, is included in AASHTO LRFD equations. This is excluded here for clarity. As a reminder of the importance of resistance factors, they are included with V_c and V_s in the equations in this section.

Shear strength of box and three-sided rectangular conduits per AASHTO LRFD – Most box conduits can be evaluated using the "sectional model" presented in AASHTO LRFD Article 5.7.3 (Vecchio and Collins, 1986). The basic strength equations for this model are

$$\phi_v V_c = \phi_v \mathrm{k}_u \ \beta \lambda \sqrt{f_c'}\, b\, d_v, \tag{7.30}$$

$$\phi_v V_s = \frac{\phi_v A_v f_y d_v \cot\theta}{s}, \tag{7.31}$$

where

ϕ_v = resistance factor for shear in concrete, 0.9 (AASHTO LRFD 5.5.4.2)
β = factor indicating ability of diagonally cracked concrete to transmit tension and shear (Equation 7.32)
λ = modification factor for concrete density, 1.0 for normal-weight concrete
f_c' = concrete compressive strength, ksi, MPa
k_u = Unit correction factor, 0.0316 for US units, 0.83 for SI units
b = width of section being designed in./ft, m/m
d_v = effective shear depth, the lesser of 0.9 d or 0.72 h, in., mm
d = depth from compression face to centroid of reinforcement, in., mm
h = wall thickness, in., mm
A_v = required area of transverse (stirrup) reinforcement
θ = angle of inclination of diagonal compression stresses
s = spacing of transverse reinforcement, in., mm

Equation 7.31 assumes stirrup reinforcement is at an angle of 90 degrees from the flexural reinforcement. For many years, shear design assumed that diagonal tension cracks formed at an angle of 45° (i.e., θ = 45°) and β = 2.0. AASHTO LRFD permits using this simplification in current practice; however, it is most often overly conservative and sometimes underconservative. For round and rectangular conduits without the minimum amount of transverse (stirrup) reinforcement:

$$\beta = \frac{4.8}{\left(1+750\varepsilon_s\right)} \frac{51k_u}{\left(39k_u + s_{xe}\right)}, \tag{7.32}$$

$$\theta = 29 + 3500\varepsilon_s, \tag{7.33}$$

where

ε_s = strain in reinforcement
s_{xe} = crack spacing parameter, in., mm
k_U = unit correction, 1.0 for US units, 25.4 for SI units

For non-prestressed members, the reinforcement strain is calculated with Equation 7.34. The equation assumes that half of the factored thrust reduces the tension force (the other half increases the compression stresses in the concrete) and includes the shear force:

$$\varepsilon_s = \frac{\dfrac{|M_u|}{d_v} - 0.5 N_u + V_u}{E_s A_s}, \tag{7.34}$$

where

$|M_u|$ = Absolute value of the factored moment, not less than $V_u\, d_v$, in.-k/ft, kN-m/m
N_u = factored axial force, taken positive in compression, k/ft, kN/m (The sign convention is reversed from AASHTO to be consistent with the nonrectangular design equations)
V_u = factored shear force, k/ft, kN/m
E_s = modulus of elasticity of reinforcement
A_s = area of tensile reinforcement, in^2/ft, m^2/m

The crack spacing parameter is calculated as

$$s_{xe} = s_x \frac{1.39}{\dfrac{a_g}{k_u} + 0.63}, \tag{7.35}$$

where

s_{xe} = crack spacing parameter as influenced by aggregate size, in. mm (12.0 in., 300 mm) $\leq S_{xe} \leq$ (80 in., 2,000 mm)
s_x = crack spacing parameter, in., mm
= d_v unless section has intermediate layers of reinforcement, which is unlikely in conduits
a_g = maximum aggregate size, in., mm
k_u = unit correction, 1 for US units and 25.4 for SI units

The limitation that $M_u \geq V_u\, d_v$ can be limiting for box conduits as the moments at the critical shear locations are often low. The AASHTO Manual for Bridge Engineering (AASHTO MBE) allows dropping this limit when load rating conduits. It could be dropped for the design of new box conduits as well.

Some common results of these calculations for conduits are as follows:

- For virtually all rectangular concrete conduits, s_{xe} = 12 in., (300 mm), and the second term of Equation 7.35 is equal to 1.0.

- If s_{xe} = 12 in., (300 mm), then $\beta > 2.0$ for reinforcement stress levels up to 55 ksi (380 MPa). In the top and bottom slabs of box conduits, the shear forces in slab are highest near the haunch where the reinforcement stresses are typically low, which gives a high value of β and thus shear strength.

Shear strength in nonrectangular sections per AASHTO LRFD – Nonrectangular sections, i.e., round, elliptical, and arch sections, are designed for shear strength at the critical section, where $M_{nu}/(V_u d) = 3.0$ in accordance with AASHTO LRFD Article 12.10.4.2.5 (Equation 7.36, Heger and McGrath, 1982, 1982a). While Equation 7.39 looks significantly different from the AASHTO LRFD Article 5.7.3 equation, it produces similar results.

$$V_n = k_1\, b\, d\, F_{vp}\, \sqrt{f_c'}\,(1.1 + 63\rho)\left(\frac{F_d\, F_n}{F_c}\right), \tag{7.36}$$

where

V_n = nominal shear capacity, k/ft, kN/m
M_{nu} = moment corrected for thrust effect, k/ft, kN/m
$= M_u - N_u\left[\frac{4h-d}{8}\right]$
b = design width of section, 12 in./ft, 1 m/m
d = depth from compression face of member to centroid of flexural reinforcement, in., m
F_{vp} = process and material factor, defined in Article 12.10.4.2.3 but typically taken as 1.0
k_1 = Unit correction factor, 0.0316 for US units, 0.083 for SI units
f_c' = compressive strength of concrete, ksi, kPa
F_d = correction factor for depth of section
$= 0.8 + \frac{k_2}{d} \le 1.3$
k_2 = unit correction: 1.6 for US 0.041 for SI
F_n = correction factor for compressive thrust (AASHTO has a different equation for tensile thrust)
$= 1 + \frac{N_u}{k_3 h}$
N_u = Thrust (positive in compression), k/ft. kN/m
k_3 = unit correction: 24 for US, 13,800 for SI
F_c = correction factor for curvature – use the plus sign for tension on inside reinforcement and minus sign for tension on the outside reinforcement
$= 1 \pm \frac{d}{2r}$
ρ = reinforcing ratio, ≤ 0.02
A_s = tension reinforcement, in^2/ft, m^2/m

7.5.4 Concrete conduits for jacked installations

Some tunnels are lined with concrete conduits that are jacked into place as the face of the tunnel is advanced. American Society of Civil Engineers Standards 27-17 (ASCE/CI 2017) and ASCE/CI 28-00 (ASCE/CI 2000) address the design of pipe and box conduit jacking installations, respectively. The magnitude of earth load on a jacked conduit is a function of the soil conditions, the size of the overbore (tunnel diameter minus conduit diameter), and whether the overbore is filled with grout after construction. Earth load is typically less, and often much less, than for a direct buried conduit. The two jacking conduit standards address earth load calculations.

The key element of jacking conduit design is addressing the load path of the jacking force through the joint. Concrete conduits for jacked installations are manufactured to tighter tolerances than typical concrete conduit to provide square ends suitable for uniformly transmitting the jacking forces from one section to the next. Joint packing material between adjacent conduit sections assists in providing this uniform load transmission (Figure 7.27). The use of straight bells, as shown in Figure 7.27, minimizes the bore diameter required to allow the conduit to pass and reduces the annulus around the conduit that typically requires a lubricant during jacking and is filled with grout after installation is complete. Extending the reinforcement into the bell and spigot ends of the conduit controls displacement should cracking in the joint occur.

The joint packing material can be specified in the contract documents or chosen by the contractor. ASCE/CI 2017 recommends ½ to ¾ in. (12 to 37 mm) plywood. The width of the packing material defines the area of concrete available to carry the jacking force. The packing material also allows jacked conduit to follow curved alignments, which result in nonconcentric pressure distributions on the conduit ends.

ASCE/CI 2017 uses the following guidelines for determining maximum jacking loads on concrete conduit:

- Load factors:
- $LFJ_1 = 1.5$ for jacking with uniform pressure around the circumference – all jacking conduits are designed for this condition (Figure 7.28a)

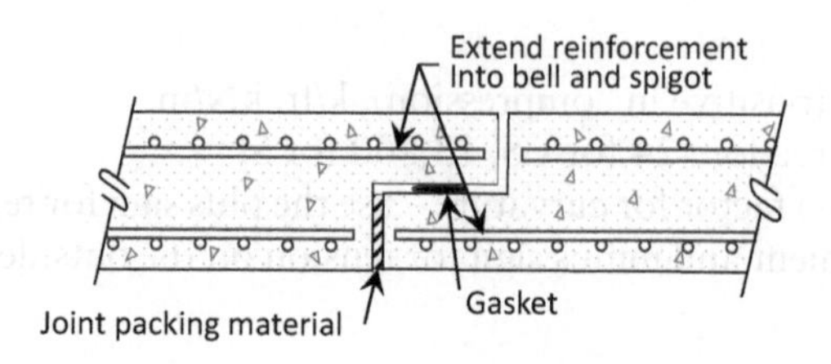

Figure 7.27 Typical joint detail for jacking conduits.

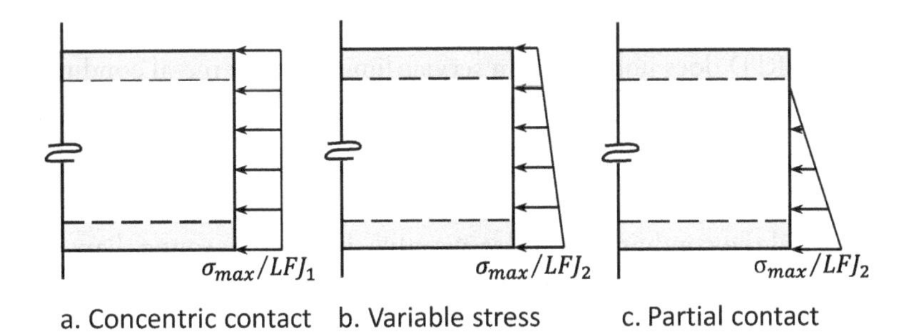

Figure 7.28 Full and partial contact jacking loads.

- LFJ_2 = 1.2 for jacking with an eccentric force causing nonuniform stress – the lower load factor is allowed for this condition on the basis that an overstress might cause a local overstress but will not result in a widespread failure (Figures 7.28b and 7.28c)
- Resistance factor: ϕ_j = 0.90.
- Maximum factored concrete stress: $\sigma_{max} = 0.85\ \phi_j f_c' / LFJ_1$.

Many designers believe that Class V conduit should always be specified for jacked installations; however, adding extra circumferential reinforcement beyond what is needed for earth load does not improve jacking capacity.

7.6 STRUCTURAL DESIGN OF METAL CONDUITS

There are several types of metal conduits, and because of the state of the art of conduit design when each type was introduced, the design methods vary widely, from totally prescriptive procedures to full finite element methods. AASHTO LRFD provides separate design guidance for four types of structures:

- Metal pipe, pipe arch, arch structures, and steel-reinforced thermoplastic conduits (Article 12.7)
- Long-span structural plate structures (Art. 12.8)
- Deep corrugated structural plate structures (Art. 12.8.9)
- Structural plate box structures (Article 12.9)

7.6.1 Metal pipe, pipe arch, arch structures, and steel-reinforced thermoplastic conduits

AASHTO LRFD Article 12.7 provides design instructions for corrugated and spiral rib metal conduits, as well as structural plate structures with seams that may be riveted, welded, lock seam, or bolted.

The primary service limit state for flexible conduits is deflection control. AASHTO LRFD does not provide a service limit state for metal conduits, as the ductility of steel and aluminum allows substantial deformations to occur. However, the AASHTO LRFD Construction Specifications (AASHTO Construction, AASHTO, 2023) require postconstruction measurements of deflection in conduits to document that construction procedures have been followed and the conduit shape is serviceable, both to prevent collapse and to prevent joints from opening and or leaking. These requirements essentially impose a service limit. Typically, the deflection limits in AASHTO Construction are a 7.5% reduction in vertical diameter or a 7.5% increase in horizontal diameter, but larger structures must also be monitored for more complex distortions such as racking, change in top plate radius and other nonuniform deformations. Further, joint performance should also be a consideration in setting allowable field deflections. Infiltration of water or soil or exfiltration of water through a joint can result in a loss of soil embedment and thus support to a conduit and, in some cases, may undermine a roadway.

Strength limit states for metal conduits include

- compression (called "wall area" in AASHTO),
- buckling strength, and
- seam resistance for structures with longitudinal joints.

Compression (wall area) – Compressive strength of conduits must be adequate to prevent yielding under the compressive thrust imposed by earth and live loads. Earth loads are assumed to be the soil prism load, and live loads are calculated in Section 7.4. As previously noted, AASHTO Article 12.7 on metal conduits references Article 12.12.3.5 on thermoplastic conduits for calculation of the term P_{FD}, the factored dead load vertical crown pressure. This is the only place in AASHTO LRFD where a load factor for earth load thrust is identified as applying to metal conduits.

$$T_L = \frac{P_{FD}(S)}{2} + \frac{P_{FL}(C_l)F_1}{2}, \tag{7.37}$$

where

T_L = factored thrust, k/ft, kN/m

P_{FD} = factored earth load, vertical crown pressure as calculated in Section 7.7.2, Equation 7.48 (LRFD Article 12.12.3.4) with *VAF* taken as 1.0 and D_o taken as *S*, k/ft², kPa

S = diameter or span, ft, m

P_{FL} = factored live load vertical crown pressure, as calculated in Section 5.3, k/ft², kPa

C_L = width of conduit on which live load is applied parallel to span, ft, m
$= l_w \leq S$
F_1 = live load modifier based on type of conduit, dimensionless
= 1.0 for thrust at the springline
= for crown thrust in corrugated metal conduit and steel-reinforced thermoplastic conduit:

$$F_1 = \frac{0.75(S)}{l_w} \geq \frac{15}{12(S)k_1} \geq 1.0 \tag{7.38a}$$

= for crown thrust in long-span corrugated metal structures:

$$F_1 = \frac{0.54(S)}{w_t + LLDF(H) + 0.03(S)} \tag{7.38b}$$

w_t = tire patch width, ft, m
H = depth of cover, ft, m
k_1 = unit correction, 1 US, 3.28 SI

Buckling strength – Buckling strength of metal conduits designed under AASHTO LRFD Article 12.7 is evaluated by an empirical equation based on a curved section of conduit restrained by the soil, *r/k*, where r is the radius of gyration of the section ($\sqrt{I/A}$) and *k* is a soil stiffness factor where a value of 0.22 is "thought to be conservative for the types of backfill material allowed for conduit and arch structures" (AASHTO LRFD, Article 12.7.2.4). Two conditions are defined:

$$\text{if } S < \left(\frac{r}{k}\right)\sqrt{\frac{24\,E_m}{F_u}}, \text{then } f_{cr} = F_u - \frac{\left(\frac{F_u\,k\,S}{r}\right)^2}{48\,E_m}; \tag{7.39}$$

$$\text{if } S > \left(\frac{r}{k}\right)\sqrt{\frac{24\,E_m}{F_u}}, \text{then } f_{cr} = \frac{12\,E_m}{\left(\frac{k\,S}{r}\right)^2}, \tag{7.40}$$

where

S = diameter or span of the conduit, in., mm
r = radius of gyration, $\sqrt{I/A}$, in., mm
k = soil stiffness factor taken as 0.22
E_m = modulus of elasticity of the conduit material, ksi, MPa
F_u = tensile strength of the conduit material, ksi, MPa
f_{cr} = critical buckling stress, ksi, MPa

I = moment of inertia of corrugated conduit section, in^4/in., mm^4/mm
A = area of conduit wall, in^2/in., mm^2/mm

If $f_{cr} < F_y$, then the required wall area of the cross-section must be recalculated using f_{cr} as the limiting stress in lieu of F_y.

A buckling design equation based on continuum theory (Moore, 1989) has been adopted into AASHTO LRFD for deep corrugated conduits (see the following). This equation is suitable and perhaps more appropriate for metal conduits than Equations 7.39 and 7.40.

Seam strength – A check on seam strength is required for metal conduits with longitudinal joints. This is a straightforward calculation:

$$SS \geq \frac{T_f}{\phi_s}, \tag{7.41}$$

where

SS = seam strength, k/ft, kN/m
T_f = total factored thrust from all applicable load conditions, k/ft, kN/m
ϕ_s = resistance factor for seam strength, 0.67

With the load factor of 1.95, the combined factor of safety, L_f/ϕ_s, is approximately 3, which is consistent with long-time practice for metal conduits.

Flexibility factor – While not considered a limit state, metal conduits also have a conduit stiffness requirement for handling and installation. This is called a flexibility factor and, as discussed in Section 3.2.1, is analogous but not identical to the conduit stiffness requirement used for thermoplastic conduits. AASHTO LRFD Article 12.5.6 provides maximum values for corrugated metal conduits based on material, corrugation shape, and thickness. The basis for the differences from product to product is not clear.

$$FF = \frac{S^2}{E_m I}, \tag{7.42}$$

where

FF = flexibility factor, in./k, mm/N
S = span of conduit, in., m
I = moment of inertial of corrugated conduit section, in.^{4}/in., m^4/m

Steel-reinforced thermoplastic conduit – Steel-reinforced thermoplastic conduit is a composite profile pipe (Section 3.1.2, Figure 7.29) designed under AASHTO LRFD Article 12.7 as a metal conduit; however, because of the unique profile, where the main structural element is a helically wound band held in place by polyethylene encasement and liner, AASHTO LRFD requires

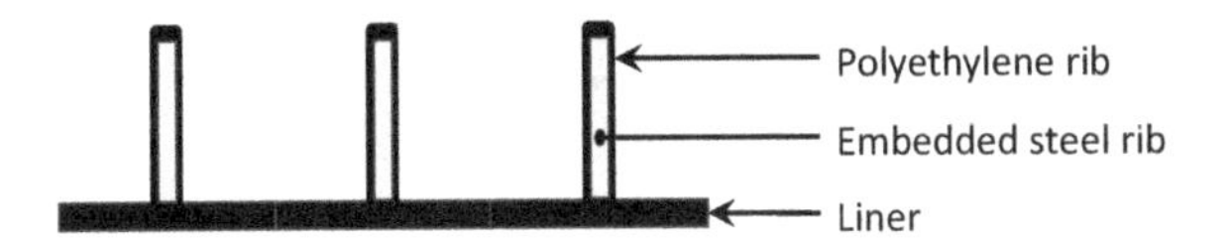

Figure 7.29 Steel-reinforced thermoplastic conduit wall profile.

that the profile performance be validated through a three-dimensional finite element analysis, full-scale testing, and stub compression testing in accordance with AASHTO Standard T 341 *Standard Method of Test for Determination of Compression Capacity for Profile Wall Plastic Pipe by Stub Compression Loading*. This standard is discussed in more detail in Section 7.7.2. Structural capacity is determined, like a corrugated metal pipe with checks for thrust and general buckling. The metal rib is assumed to be supported against local buckling by the polyethylene encasement. Like other metal conduits in this section, there is no deflection calculation during design, but AASHTO Construction requires a deflection check after installation and limits the deflection change to 5% of the diameter.

7.6.2 Long-span structural plate structures

Long-span structural plate structures are designed under AASHTO LRFD Article 12.8. The name of the category of conduit sounds generic in nature; however, in AASHTO, LRFD it references a specific structure class, which includes the shapes presented in Figure 7.30.

Long-span structures were developed prior to the general availability of finite element analysis and are sensitive to deflection and distortion during construction. The introduction of these structures was a prime motivation for developing the computer program CANDE (CANDE 2022). Because of the lack of suitable analytical methods and design procedures, these conduits are designed prescriptively, without load or resistance factors. Design is completed by the manufacturer, and AASHTO Construction requires a qualified manufacturer's representative to serve as the "shape-control Inspector" during construction. With this inspection, these long-span structures have provided good service.

The primary elements of long-span structures are made up of 6 in. x 2 in. (150 mm x 50 mm) corrugated structural plates. However, to provide extra stiffness during backfilling, these structures are required to have "special features," of which there are two types:

- Longitudinal stiffeners – Longitudinal stiffeners, often known as thrust beams, consist of reinforced concrete "beams" cast along the length of the structure (Figure 7.31a). Longitudinal stiffeners contribute minimally to resisting earth and live loads; however, during backfilling, they distribute the localized compaction forces along the length of the

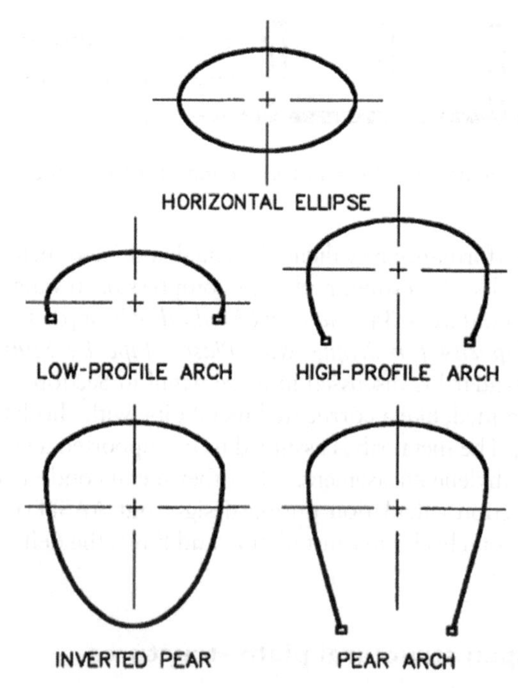

Figure 7.30 Long-span structural plate conduit shapes. (From AASHTO LRFD Bridge Design Specifications, 2020, published by the American Association of State Highway and Transportation Officials, Washington, DC. Used with permission)

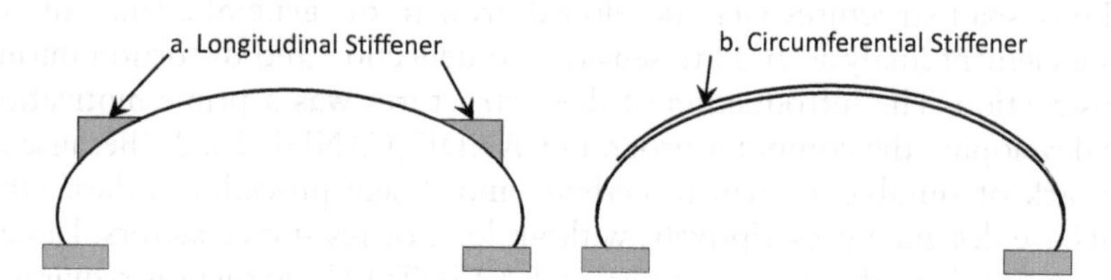

Figure 7.31 Long-span stiffener types.

conduit, effectively mobilizing a greater length of structure to resist upward peaking and distortion.

- Circumferential stiffeners – Circumferential stiffeners are additional structural elements bolted to the top arc of a long-span conduit (Figure 7.31b). This element stiffens the top arc to offer greater resistance to compaction forces during backfilling as well as to earth and live loads after construction.

AASHTO LRFD provides no specific requirements for the size, shape, or stiffness of either type of stiffener. They are designed by the manufacturer.

The design of long-span structures is carried out by the manufacturer, based on AASHTO LRFD Table 12.8.3.1.1-1 (Table 7.10), which determines the plate thickness and minimum depth of fill based on the radius of the top plates and sets other geometric restrictions. Checks for thrust and seam strength are required, but the requirements for buckling and flexibility factor are waived for long-span conduits. There is evidence that buckling can occur (Moore, R.G. et al., 1995). Earth load thrust is calculated using twice the top arc radius rather than the span. For shapes such as arches that have small radius corners, it is important to provide good soil support. The thrust is reasonably constant at these locations, but the interface soil stress, $p = T/R$, where T is the calculated thrust and R is the mean radius at the location being considered, increases significantly at these locations. The soil stiffness must be adequate to support the higher pressures.

A table presents top arc minimum thickness for a 6 inch by 2 inch corrugated steel plate across various top radii, with corresponding geometric limits and minimum cover requirements.

AASHTO LRFD Article 12.8.7 notes that concrete relieving slabs may be used to reduce moments in long-span structures, but as with the rest of the design method, it provides no design guidance on the benefit that is derived.

Table 7.10 AASHTO Design Requirements for Long Span Structures with Special Features

Top Arc Minimum Thickness (in.)					
Top Radius (ft)	≤15.0	15.0–17.0	17.0–20.0	20.0–23.0	23.0–25.0
6″ × 2″ Corrugated Steel Plate—Top Arc Minimum Thickness (in.)	0.111	0.140	0.170	0.218	0.249

Geometric Limits

The following geometric limits shall apply:

- Maximum plate radius—25.0 ft
- Maximum central angle of top arc—80.0°
- Minimum ratio, top arc radius to side arc radius—2
- Maximum ratio, top arc radius to side arc radius—5

Minimum Cover (ft)					
Top Radius (ft)	≤15.0	15.0–17.0	17.0–20.0	20.0–23.0	23.0–25.0
Steel thickness without ribs (in.)					
0.111	2.5	—	—	—	—
0.140	2.5	3.0	—	—	—
0.170	2.5	3.0	3.0	—	—
0.188	2.5	3.0	3.0	—	—
0.218	2.0	2.5	2.5	3.0	—
0.249	2.0	2.0	2.5	3.0	4.0
0.280	2.0	2.0	2.5	3.0	4.0

Source: From AASHTO LRFD Bridge Design Specifications, 2020, published by the American Association of State Highway and Transportation Officials, Washington, D.C. Used with permission

McGrath et al. (2002) studied the design of long-span metal conduits with the intent of developing more rigorous design procedures, and Liu et al. (2024) have recently produced equations for estimating the live load moment based on three-dimensional finite element analyses. However, making realistic predictions of bending moments, even with finite element models, is difficult because of the dependence of moment on backfill materials and procedures. The final proposal recommended setting allowable deformations and calculating the bending moment based on the associated change in curvature of the elements. AASHTO did not adopt the recommendations and continues to use the prescriptive method of design.

Long-span structures were unique at the time of their development because of the large spans and low structural stiffness. To address this uniqueness, AASHTO LRFD Article 12.8 contains several sections to address design, backfill, and construction issues that might be considered routine design matters with the aid of finite element analysis. These include:

- differential settlement of footings, including side-to-side and longitudinal – both of which need to be addressed through site investigation prior to construction;
- distribution of live load to footings – allowing the load to spread longitudinally below the crown of the structure;
- embedment design – consideration of lateral soil stresses imposed on backfill by the structure and on the native soils by the backfill;
- end treatment – addressing the support required if the end of the structure is a sloping embankment or a headwall that may be skewed; and
- hydraulic considerations – i.e., the need for cut-off walls and design for hydraulic uplift and scour.

Katona and Akl (1987) proposed that the thrust force in deeply buried long spans and other conduits assembled by bolting corrugated structural plates together could be reduced by introducing slotted holes at the bolted joints rather than standard circular holes. The plates are assembled with the bolts not fully tightened, allowing slippage and circumferential shortening as earth load is added. This circumferential shortening results in the arching of load off the structure and into the sidefill, thus increasing the allowable depth of fill. The slots and sliding joints mimic a low circumferential modulus of elasticity. Both the Burns and Richard elasticity theory (Burns and Richard, 1964, see Section 5.1) and finite element analysis can be used to demonstrate this effect. Corrugated polyethylene conduits display the same behavior by virtue of a low cross-sectional wall area and a low long-term modulus of elasticity (see Section 7.7.2)

7.6.3 Deep corrugated structural plate structures

Corrugated structural plates with deeper corrugations were developed after finite element analysis was fully developed as a design tool for buried structures. AASHTO LRFD uses the "deep corrugated" classification for structural plate structures with corrugation depths greater than 5 in. (125 mm). The deeper section provides more stiffness and allows for a more rigorous approach to design. The deep corrugated profiles included in AASHTO LRFD are 15 in. x 5-1/2 in. (381 mm x 140 mm) and 20 in. x 9.5 in. (500 mm x 240 mm), but slightly different versions are available in the marketplace. AASHTO LRFD Article 12.8.9 provides guidance for the design of these structures. Unlike long-span structures, deep corrugated structures are designed for flexure and buckling.

AASHTO LRFD considers two types of deep corrugated structures: those with a ratio of crown-to-haunch radius less than 5.0 and those with a ratio of crown-to-haunch radius equal to or greater than 5.0. The shapes with higher ratios are considered box structures, and those with higher ratios are considered pipe or arch structures. Figure 7.32 shows two types of structures to demonstrate the significance of a high ratio crown-to-haunch radius.

Box structures (high ratio) resist loads predominantly through flexure and have peak positive moments at the midspan and peak negative moments at the haunch (some box structures are designed under AASHTO LRFD Article 12.9; see Section 7.6.4). Arch and other shapes (low ratio) also develop bending moments but thrust forces are significant and in some shapes dominate the response. The two types of structures have different requirements for width of structural backfill, but both are dependent on soil support to provide good performance.

Finite element analysis is used to assess soil structure interaction of deep corrugated structures to determine moment and thrust forces for design. AASHTO LRFD requires that the vertical arching factor (*VAF*) for thrust at the springline is not less than 1.3.

Minimum depth of cover varies with the type of structure. For deep corrugated structures with the ratio of crown radius to haunch radius ≥ 5.0, the minimum depth of cover is 3.0 ft (0.9 m) or the minimum depth for long-span conduits based on top radius and plate thickness (Table 7.10). For deep corrugated structures with the ratio of crown radius to haunch radius < 5.0, the minimum depth of cover is 1.5 ft (0.5 m) for spans ≤ 25 ft, 5 in. (7.75 m), and 2.0 ft (0.6 m) for larger spans.

Because of the higher stiffness of deep corrugated structures, both moment and thrust forces may be significant, and an interaction curve is required to

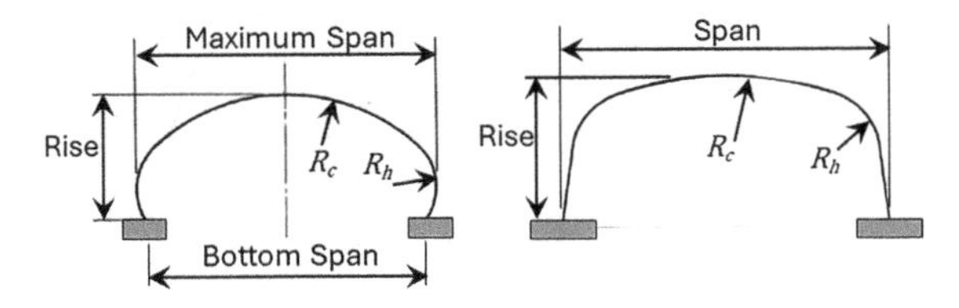

Figure 7.32 Deep corrugated structures with low (arch) and high (box) ratio of crown-to-haunch radius.

determine capacity. The interaction equation used is taken from the Canadian Highway Bridge Design Code, CSA S6. The shape of the curve is shown in Figure 7.33:

$$\left(\frac{T_f}{R_t}\right)^2 + \left|\frac{M_u}{M_n}\right| \leq 1.0, \tag{7.43}$$

where

T_f = factored thrust, k/ft, kN/m
R_t = factored thrust resistance, k/ft, kN/m
= $\phi_b F_y A$
F_y = steel yield strength, ksi, kPa
M_u = factored applied moment, k-ft/ft, kN-m/m
M_n = factored moment resistance, k-ft/ft, kN-m/m
= $\phi_b M_p$
= resistance factor for formation of a plastic hinge
M_p = plastic moment capacity of section, k-ft/ft, kN-m/m

Deep corrugated structures are manufactured as independent plates bolted together in the field. Because the structures are designed for flexural capacity, these structures have a flexural requirement for longitudinal connections. The factored moment resistance must be greater than the factored applied moment but not less than 75% of the factored moment resistance of the members being connected or the average of the factored applied moment and the factored moment resistance of the members being connected.

Deep corrugated structures use a buckling equation (Moore and Selig, 1990) that was recommended for all long-span conduits by McGrath et al. (2002). The equation is based on continuum theory (see Moore, 1989), which captures the stiffness of the embedment soil more effectively than spring models and provides a rational method of addressing the effect of shallow cover on soil support. The equation is applicable to all metal

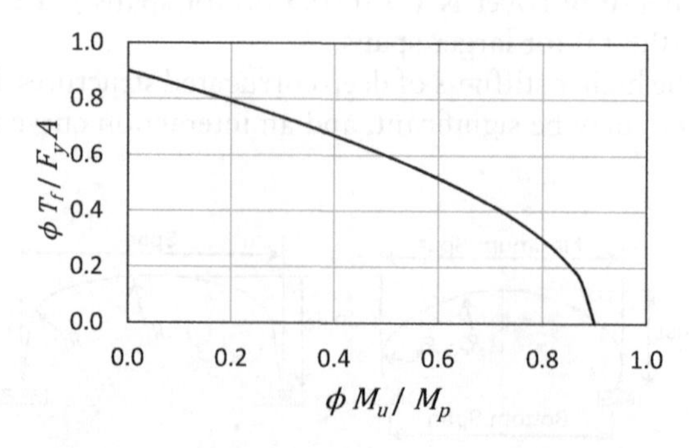

Figure 7.33 Interaction curve for combined thrust and moment of deep corrugated structures.

conduits and has been adopted for plastic conduits but expressed in terms of limiting strain (see Section 7.7.2). Liu et al. (2024) present additional methods to analyze the buckling of metal conduits under live loads.

$$R_b = 1.2\phi_b C_n \left(E_p I_p\right)^{\frac{1}{3}} \left(\phi_s M_s k_b\right)^{\frac{2}{3}} R_h, \tag{7.44}$$

where

R_b = nominal axial force in conduit wall to cause general buckling, lb/in., kN/m
ϕ_b = resistance factor for general buckling
C_n = 0.55, scalar calibration factor to account for some nonlinear effects
E_p = modulus of elasticity of conduit wall material, ksi, KPa
I_p = moment of inertia of conduit wall, including stiffeners if used, in.^{4}/in., m^4/m
ϕ_s = resistance factor for soil
M_s = Constrained modulus of embedment based on free field vertical soil stress at a depth halfway between crown and springline, ksi, KPa
k_b = Soil stiffness correction factor

$$= (1-2\nu)/(1-\nu^2)$$

ν = Poisson's ratio of soil
R_h = 11.4/ (11 + S /H), correction factor for backfill geometry
S = Conduit maximum span, ft, m
H = Depth of fill over top of conduit, ft, m

The equation shows that the key parameters in developing buckling resistance are the conduit stiffness (E_p I_p) and soil stiffness (M_s). Since the soil stiffness is raised to a higher exponential power, increasing soil stiffness around a conduit increases buckling resistance more effectively than increasing the conduit stiffness.

7.6.4 Structural plate box structures

Structural plate box structures in AASHTO LRFD Article 12.9 are a specific set of structures for which a semi-empirical design method was developed. These boxes are manufactured with aluminum or steel corrugated structural plate reinforced with stiffeners, in particular at the haunches. Like long-span structures, the designs for this set of box conduits were developed prior to general acceptance of finite element methods for designing buried structures, and thus a simplified method of hand calculation was developed based on finite element analysis and full-scale field testing. However, like long spans, the assumptions made to create the design method required restricting certain parameters of the structures to be designed. The geometry of structural plate box structures is defined in Figure 7.34, and the limits on geometry are presented in Tables 7.11 and 7.12 for spans in the two size ranges in the specifications. Allowable depths of fill for these structures are 1.4 ft to 5.0 ft (0.43 m to 1.5 m).

Commentary to AASHTO LRFD states that these structures are controlled by flexure alone, and thrust may be ignored. This is possible because of the low cover heights, a restriction not placed on deep corrugated structures. Loads are not specifically calculated, as the design equations focus on determining the total moment, which is the sum of the positive moment at midspan and the absolute value of the negative moment in the haunch. Separate equations are provided for earth load and live load. AASHTO LRFD allows some leeway in apportioning the total moment to the crown and haunch region. This is because a failure would require plastic hinges to form at both locations.

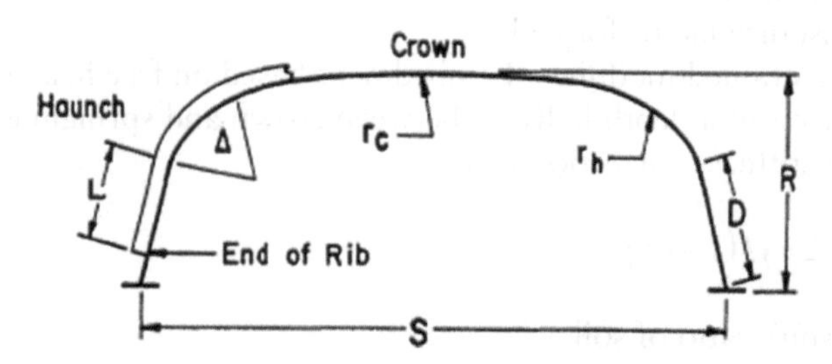

Figure 7.34 Geometry of structural metal plate box conduits. (From AASHTO LRFD Bridge Design Specifications, 2020, published by the American Association of State Highway and Transportation Officials, Washington, DC. Used with permission)

Table 7.11 Geometric requirements for box spans from 8 ft 9 in. to 25 ft 5 in. (2.67 m to 7.75 m)

Parameter	*US Units*	*SI Units*
Span, *S*	8 ft 9 in. to 25 ft 5 in.	2.67 m to 7.75 m
Rise, *R*	2 ft 6 in. to 10 ft 6 in.	0.76 m to 3.20 m
Radius of Crown, r_c	≤ 24 ft 9 ½ in.	7.56 m
Radius of haunch, r_h	≥ 2 ft 6 in.	0.76 m
Haunch radius included angle, Δ	48° to 68°	
Length of leg, *D* measured to the bottom of the plate	4 ¾ in. to 71 in.	120 to 1800 mm
Minimum length of rib on leg, *L*	Min. (19.0 in., *D*-3.0 in.) Or within 3 in. of the top of the footing	Min (480 mm, *D* – 75 mm) Or within 75 mm of the top of the footing

Source: From AASHTO LRFD Bridge Design Specifications, 2020, published by the American Association of State Highway and Transportation Officials, Washington, DC. Used with permission.

Table 7.12 Geometric requirements for box spans from 25 ft 6 in. to 36 ft 0 in. (7.77 m to 10.97 m)

Parameter	*US Units*	*SI Units*
Span, S	25 ft 6 in. to 36 ft 0 in.	7.77 m to 10.97 m
Rise, R	5 ft 7 in. to 14 ft 0 in.	1.70 m to 4.27 m
Radius of Crown, r_c	≤ 26 ft 4 in.	8.03 m
Radius of haunch, r_h	≥ 3 ft 8 in.	1.12 m
Haunch radius included angle, Δ	48° to 68°	
Length of leg, D measured to the bottom of the plate	4 ¾ in. to 71 in.	120 to 1800 mm
Minimum length of rib on leg, L	Min. (28.0 in., D-3.0 in.) Or to within 3 in. of the top of the footing	Min (710 mm, D – 75 mm) Or to within 75 mm of the top of the footing

Source: From AASHTO LRFD Bridge Design Specifications, 2020, published by the American Association of State Highway and Transportation Officials, Washington, DC. Used with permission.

The specifications provide an equation to determine the thickness of a concrete relieving slab, which can be used to spread live load out over a wider area, reducing the imposed moments in the structure. However, the AASHTO LRFD Commentary associated with the method indicates that finite element analysis is required to determine the benefit of the slab in reducing the conduit response. The method is based on the embedment soil between the structure and the slab, the axle load, and the concrete strength (Duncan et al., 1985). However, the method does not mention the need for reinforcement of the relieving slab and does not present a method to determine the reduction in moments in the structure that are achieved by the relieving slab.

7.7 PLASTIC CONDUITS

Plastic conduits do not have a long history of concrete and metal conduits for use in gravity flow applications but are well established. The first report on the subject from AASHTO (Chambers et al. 1980) was undertaken when thermoplastic conduits were small, 4 to 16 in. (100–400 mm) diameter, and the interest was applications for underdrains and side drains. Corrugated polyethylene conduits were widely used in nontransportation drainage applications prior to that time. Fiberglass conduit was available in larger sizes but was not used in transportation applications. Plastic conduits are flexible conduits and the industry looked to design procedures used for corrugated metal conduits when initially developing design methods for thermoplastics.

AASHTO LRFD has two categories of plastic conduits – thermoplastic (Article 12.12) and fiberglass (Article 12.15).

7.7.1 Flexibility and limit states

Thermoplastic gravity flow conduits are designed in AASHTO LRFD based on strain rather than stress to address the effects of the low modulus of elasticity and the high ratio of short- to long-term modulus of elasticity for some thermoplastics. The design equations have more terms than other materials to address the effects of the low modulus as circumferential shortening due to earth load thrust and external water, if present, increase the deflection above that due to bending.

Plastic conduits are required to meet a flexibility limit as were metal conduits, but all types of plastic conduits are held to the same standard in AASHTO LRFD:

$$FF = \frac{D_m^2}{E_p I_p} \leq 95\, in./k, 0.54\,\text{mm}/\text{N}, \tag{7.45}$$

where

FF = Flexibility factor, in./k, m/kN
E_p = Modulus of elasticity of conduit material, ksi, kPa
I_p = Moment of inertia, in.^{4}/in., m^4/m
D_m = Diameter of conduit to the centroid of the conduit wall, in., m

Service limit state – deflection and strain: Thermoplastic and fiberglass conduits include a limit state for deflection. Thermoplastic conduits also have a limit on service tensile strain, which will be addressed next, where factored strain levels are evaluated.

Deflection is calculated using Spangler's Iowa formula (Section 5.2.1) with an extra term to address the contribution of compression strain associated with circumferential thrust.

$$\frac{\Delta_y}{D_m} = \frac{K_B\left(D_l\, P_{sp} + C_l\, P_L\right)}{\dfrac{E_p I_p}{R_m^3} + 0.061 M_s} + \varepsilon_{sc}, \tag{7.46}$$

where

Δ_y/D_m = change in vertical diameter expressed as a fraction of the diameter to the centroid of the conduit wall
D_m, D_o = diameter to the centroid of the conduit wall and outside diameter, in., m
K_B = bedding coefficient
D_l = deflection lag factor
P_{sp} = soil prism load, psi, kPa

$$= \gamma_s (H + 0.11 D_o)$$

γ_s = density of backfill over conduit, lb/in.^{3}, kN/m^3
H = depth of fill to top of conduit, in., m
P_L = live load pressure, psi, kPa
C_L = live load coefficient for spread of live load at the crown of the conduit $= l_w/D_o \leq 1.0$ (see Section 5.3 and Figure 5.14)
l_w = distribution width of live load at the top of the conduit, in., m
W_L = live load, psi, kPa
R_m = radius to the centroid of the conduit wall, in., m
M_s = soil stiffness parameter, constrained modulus of soil (Table 4.6) or Howard's E' (Table 4.5, and Equation 7.47 for trench installations), psi, kPa
ε_{sc} = compression strain in conduit wall due to hoop thrust, in./in., mm/mm (see Section 7.7.2)

Notes on the use of the Iowa formula include the following:

- Equation 7.46 is used to provide a guide that keeping the field deflection under the allowable is feasible. The stiffness term in the denominator of the Iowa formula, 0.061 M_s, is significantly larger than the conduit stiffness term, $E_p I_p / R_m{}^3$, for most flexible thermoplastic conduits in reasonably stiff soil (Figure 7.35), meaning the conduit response is controlled by the soil. This emphasizes the need to use good construction practices to control deflection levels.
- When conduits are laid in narrow trenches, the soil stiffness of the in situ soils can affect the soil support to the conduit. AWWA Manual M45 (AWWA, 2014) provides a table of modification factors to decrease or increase the value of the soil stiffness for the deflection equation.

$$M_s = S_c M_{sb}, \tag{7.47}$$

where

S_c = Correction factor to account for stiffness of the native material in trench wall, expressed as a function of M_{sn} / M_{sb} (Table 7.13)
M_{sb} = Constrained modulus of backfill, psi, kPa
M_{sn} = Constrained modulus of trench wall material, psi, kPa (Table 7.14)
B_d = Trench width, ft, m
D_o = Outside diameter of conduit, ft, m

Table 7.14 is condensed from Manual M45, but interpolation for intermediate values will yield essentially the same results. This table is

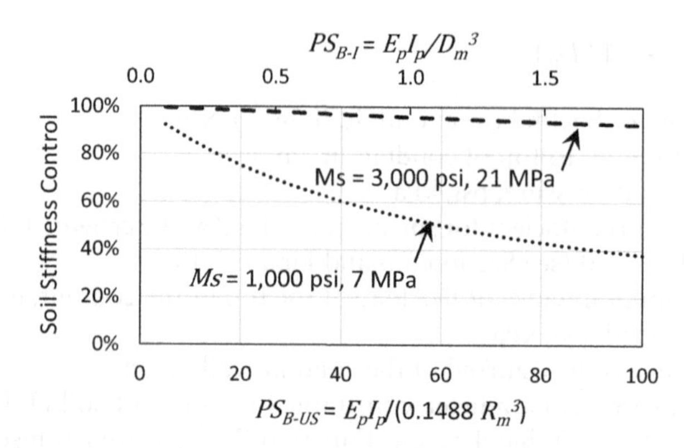

Figure 7.35 Soil stiffness contribution to resisting deflection.

Table 7.13 Combining factor S_c to account for native soil stiffness

M_{sn}/M_{sb}	B_d/D_o					
	1.25	1.5	2	2.5	3	5
	S_c					
0.05	0.10	0.15	0.27	0.38	0.58	1.00
0.2	0.25	0.30	0.47	0.58	0.75	1.00
0.4	0.45	0.50	0.64	0.75	0.85	1.00
0.8	0.84	0.87	0.93	0.96	0.98	1.00
1	1.00	1.00	1.00	1.00	1.00	1.00
2	1.70	1.50	1.30	1.20	1.11	1.00
5	3.00	2.20	1.70	1.50	1.30	1.00

Source: Condensed from AWWA, 2014, reprinted with permission. M45 Fiberglass Pipe Design, 3rd edition, American Waterworks Assn.

based in part on the assumption that the in situ soil will have no effect on conduit response if the trench width is 5 D_o or greater.

- The bedding factor proposed by Spangler varied from 0.083 to 0.110 as a function of the bedding angle and addressed how bedding geometry influenced the change in horizontal diameter. As noted in Section 5.2.1, the bedding factor is unlikely to represent the effect of bedding support on the vertical diameter. Howard's calibrations of *E'* (Section 4.3.1) used a bedding factor of 0.1 for all field installations in his data set. It is common in practice to do the same in design.
- Spangler proposed using the soil prism load as the input to the Iowa formula based on the assumption at the time that this was the best estimate of the earth load on flexible conduits. Howard's calibration

Table 7.14 Constrained modulus estimates for in situ soils adjacent to embedment zone

Granular[d]		*Cohesive*			M_{sn}[a]	
		q_u[c]				
Blows/ft[b] *(0.3 m)*	*Description*	*Tons/sq. ft*	*kPa*	*Description*	*psi*	*MPa*
> 0–1	very, very loose	> 0–0.125	0–13	very, very soft	50	0.35
1–2	very loose	0.125–0.25	13–25	very soft	200	1.4
2–4		0.25–0.5	25–50	soft	700	4.8
4–8	Loose	0.50–1.0	50–100	medium	1,500	10.3
8–15	slightly compact	1.0–2.0	100–200	stiff	3,000	20.7
15–30	compact	2.0–4.0	200–400	very stiff	5,000	34.5
30–50	dense	4.0–6.0	400–600	hard	10,000	69
> 50	very dense	> 6.0	> 600	very hard	2,000	138

[a] M_{sn} for rock is ≥ 50,000 psi, (345 MPa)
[b] Standard penetration test per ASTM 1586
[c] q_u is the unconfined shear strength
[d] See AWWA M45 for additional notes

Source: AWWA, 2014, reprinted with permission. M45 Fiberglass Pipe Design, 3rd edition, American Waterworks Assn.

for E' uses the same assumption; thus, even though the thrust force at the springline, a common measure of earth load, can be greater or less than the soil prism weight, the soil prism load is the appropriate input for the Iowa formula.

- The live load coefficient, C_L, considers conduits with low cover such that the load distribution in the soil (l_w Figure 5.14) has not spread to the full diameter of the conduit.
- The hoop strain, ε_{sc}, represents the circumferential shortening (and hence, reduction in diameter) of the conduit due to earth load. For some conduits with low modulus of elasticity and low cross-sectional wall area, this shortening can be significant, especially as the depth of fill increases. This is addressed next.

The key item about predicting deflection is that the calculation gives an *estimate* of the change in diameter that is useful for evaluating backfill types and compaction levels for a specific project. The actual control of deflections is achieved in the field using specified materials and good construction procedures.

McGrath and Chambers (1981) published deflection data on a series of conduits installed as part of a field test (Figure 7.36). The test conduits represented a range of stiffnesses ($PS_{B\text{-}US}$ = 18 – 230 psi, $PS_{B\text{-}I}$ = 2.3 – 29.5 kPa) in a soil classified as SP-SM per ASTM D 2387, or Class III per ASTM D 2321, and compacted to 92-95% of maximum standard Proctor density.

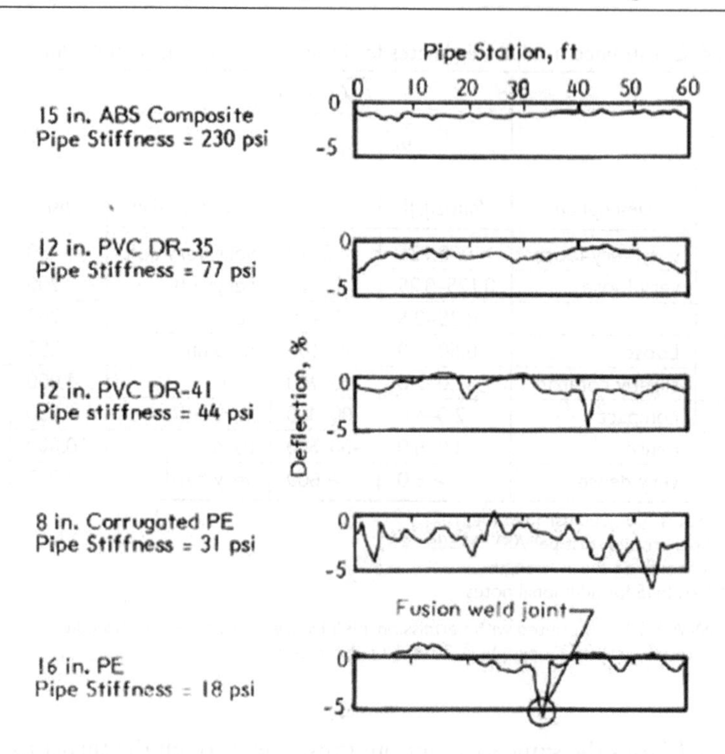

Figure 7.36 Deflection of conduits installed in compacted sand. (McGrath and Chambers, 1981, with permission from ASCE)

Notes: 1 in.= 25.4 mm, 1 ft = 0.3 m, 1 psi = 6.89 kPa; Pipe stiffness is $PS_{B\text{-}US}$

The final cover depth was 20 ft (6 m). Although the test conduits were small in diameter by today's standards, the data shows several trends:

- The variability of the deflection profiles increases with decreasing conduit stiffness. The ABS conduit shows very little longitudinal variation.
- The corrugated PE conduit, which did not have a liner to provide better hydraulic flow, had the most variability due to the low hoop and longitudinal stiffness. Lined corrugated PE conduits, which are in common use today, have higher longitudinal stiffness.
- The 16 in. (400 mm) diameter PE conduit had a solid wall and was the lowest stiffness conduit. It showed increased variability relative to the noncorrugated conduits, but the average deflection was less than 1%. Due to the low stiffness, compaction of the backfill resulted in upward deflection (i.e., vertical diameter increase). Flexible conduits with low stiffness can deflect upward excessively and distort during backfilling if the compaction forces are excessive.

The minimum conduit stiffness required to withstand the installation process without excessive deflection or distortion has been much debated.

Metal conduits have long relied on the flexibility factor ($D_i^2/(E\ I)$) to determine minimum conduit stiffness. This parameter allows the conduit stiffness ($EI/(0.1488\ R^3)$ or EI/D^3) to decrease with increasing diameter (Figure 3.26). Many plastic conduit standards have used $PS_{B\text{-}US}$ = 46 psi (320 kPa) as a minimum stiffness. The metal conduit approach is reasonable as larger diameter/span conduits tend to be heavier and more resistant to deformation; however, in the extreme case of long-span metal conduits (Section 7.6.2), full-time field monitoring is required to ensure that deformations are controlled. Section 9.3 describes an installation of a fiberglass conduit where extreme compaction forces created significant distortions in a low-stiffness conduit. AWWA Manual M45 (AWWA, 2014) allows the use of conduit with stiffness $P_{Sb\text{-}US}$ 9 psi ($PS_{B\text{-}I}$ 370 kPa) but restricts the allowable backfills to coarse-grained materials (SC1 and SC2).

The backfill type and compaction methods contribute to the choice of minimum conduit stiffness. Coarse-grained soils require modest compactive effort to reach a desired unit weight, which is desirable, but can still distort under high compactive energy. Soils with increasing quantities of fines require increasingly more energy to compact (see Sections 4.3.4 and 5.4.2). Flowable fill requires no compactive effort, which facilitates the use of low-stiffness conduits. Similarly, uniform rounded backfill such as pea gravel can be placed with little or no compactive effort.

One additional point flowing from the data presented in Figure 7.36 is that a conduit can fail deflection testing because of the maximum deflection, so control of variability in backfill and compaction procedures is important.

Strength limit states - Strength limit states for plastic conduits include the following:

- Compression (called Wall area in AASHTO LRFD) Flexure (fiberglass)
- Buckling
- Flexibility limit

Table 7.15 presents the resistance factors for plastic conduits in AASHTO LRFD:

Table 7.15 Resistance factors for plastic conduits

Thermoplastic Conduits	
Thrust, ϕ_T	1.00
Soil stiffness, ϕ_s	0.9
Global buckling, ϕ_{bck}	0.7
Flexure, ϕ_f	1.00
Fiberglass Conduits	
Flexure, ϕ_f	0.9
Global buckling, ϕ_{bck}	0.63

The resistance factors for plastic conduits, like other conduits in AASHTO LRFD, have not been calibrated to incorporate actual variability. Buckling of plastic conduits, like other types of conduits, is known to be variable, and thus the resistance factor for buckling is set to a lower value,

7.7.2 Thermoplastic conduits

Thermoplastic conduits included in AASHTO LRFD include corrugated, solid wall, and profile wall polyethylene (PE), solid wall and profile wall polyvinyl chloride (PVC), and corrugated polypropylene (PP). Examples of profile walls for PE and PVC are shown in Figures 7.37a and 7.37b, respectively. Not all these profiles are currently available. AASHTO LRFD Table 12.12.3.3-1 provides required plastic resin cell classifications, and the key material properties and strain limits associated with each type of product. Some products are manufactured with pressure-rated resins, but pressure rating is not required for gravity flow conduits, which are in a stress relaxation condition after installation. The table also provides the product standards, which should be reviewed for additional details. Table 7.16 presents key properties from the AASHTO LRFD table.

Table 7.16 provides a compression strain limit for thermoplastics. Although not listed here, AASHTO LRFD allows a 50% increase in the compression strain limit when evaluating combined flexure and compression strains.

Thrust – Thrust in plastic conduits is the result of soil overburden, vehicle loads, and external water loads. External water load can increase the compression strain in corrugated PE and PP conduits because of the low cross-sectional area. The load modifiers for vertical earth, η_{EV}, and live load, η_{LL}, are also listed in the equation as a reminder that the values differ, as live load is considered redundant because of its localized nature – that is, if

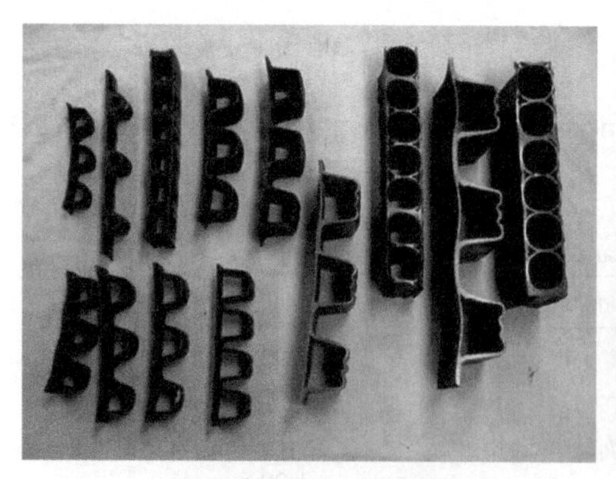

a. Profile PE Conduit Walls

b. Profile PVC Conduit Walls

Figure 7.37 Examples of profile walls for thermoplastic conduits.

Table 7.16 AASHTO material properties for thermoplastic conduits

Products	Service Long-Term Tension Strain Limit, ε_{yt} (%)	Factored Compression Strain Limit, ε_{yc} (%)	Initial Properties		75 Year Properties	
			F_u min (ksi)	E min (ksi)	F_u min (ksi)	E min (ksi)
Solid/Profile Wall PE	5.0	4.1	3.0	110	1.4	21
Profile Wall PE	5.0	4.1	3.0	80	1.1	19
Corrugated PE	5.0	4.1	3.0	110	0.9	21
Solid/Profile Wall PVC	5.0	2.6	7.0	400	3.6	137
	3.5	2.6	6.0	440	2.5	156
Corrugated PP	2.5	3.7	3.5	175	1.0	28

Note: This table is consolidated from AASHTO LRFD. Users must check the complete table for design or specification purposes.

Source: From AASHTO LRFD Bridge Design Specifications, 2020, published by the American Association of State Highway and Transportation Officials, Washington, DC. Used with permission.

overloaded, the live load can spread over a greater longitudinal distance to carry the load. Earth load is considered nonredundant as the load is uniform along the length of the conduit. Factored thrust is calculated as

$$T_u = \left[\eta_{EV}\left(\gamma_{EV}\, K_{\gamma E}\, K_2\, VAF\, P_{sp} + \gamma_{WA}\, P_W\right) + \eta_{LL}\,\gamma_{LL}\, P_L\, C_L\, F_1\, F_2\right]\frac{D_o}{2}, \quad (7.48)$$

where

T_u = Factored thrust, k/in., kN/m
$K_{\gamma e}$ = 1.0, Installation factor (see the following discussion)
K_2 = Coefficient based on conduit location, 1.0 for springline, 0.6 for crown
VAF = Vertical arching factor

$$= 0.76 - 0.71\left(\frac{S_H - 1.17}{S_H + 2.92}\right)$$

S_H = Hoop stiffness ratio (see Section 5.1.1)

$$= \frac{\phi_s\, M_s\ R_m}{E_p A_p}$$

M_s = Constrained modulus of soil, ksi, MPa
R_m = Radius to centroid of conduit wall, in., m
E_p, A_p = Modulus of elasticity and cross-sectional area per unit length of conduit wall
P_{sp}, P_{WA}, P_L = Soil prism pressure, live load pressure and hydrostatic pressure, ksi, kPa
η_{EV}, η_{LL} = Load modifiers (Section 7.4.1)
γ_{EV}, γ_{WA}, γ_{LL} = Load factors (Section 7.3)
C_L = Live load coefficient to limit width of live load pressure over conduit, $L_w/D_o \leq 1.0$
D_o = Outside diameter of conduit, in., m
F_1 = Live load coefficient applied to metal and plastic conduits

$$= \frac{0.75(S)}{l_w} \geq \frac{15}{12(D_i)k_1} \geq 1.0$$

D_i = Insided diameter of conduit, in., m
k_1 = Unit correction, 1.0 US, 39
F_2 = Live load coefficients for plastic conduits

$$= \frac{0.95}{1+0.6S_H}$$

Compression in the conduit wall is limited by strain:

$$\varepsilon_{uc} = \left[\frac{\eta_{EV}\left(\gamma_{EV}\,K_{EV}\,VAF\,P_{sp} + \gamma_{WA}\,P_W\right)}{A_{eff}\,E_{50}} + \frac{\eta_{LL}\,\gamma_{LL}\,P_L\,C_L\,F_1\,F_2}{A_{eff}\,E_o}\right]\frac{D_o}{2}, \quad (7.49)$$

where

ε_{uc} = factored compression strain due to thrust
A_{eff} = effective area of the conduit wall per unit length based on local buckling, in.^{2}/in., mm^2/mm
E_o = short-term modulus of conduit material, psi, kPa
E_{50} = long-term modulus of conduit material, psi, kPa

Service load thrust strain, ε_{sc}, is computed by setting the load factors and load modifiers in Equation 7.49 equal to 1.0. The service load thrust strain is used in the deflection calculation (Equation 7.46). As noted, the hoop stiffness factor, S_H, is high for some thermoplastics, resulting in significant hoop strain and an associated reduction in conduit diameter under thrust loading. This results in the arching of the load off the conduit, particularly for corrugated PE and PP conduits. Similar to the behavior of corrugated structural plate conduits with slotted joints (Section 7.6.2).

As currently written, the installation factor, K_{EV}, in the AASHTO LRFD 9th edition thrust equation should be taken as one. McGrath et al. (2009) recommended that the load factor for thrust be reduced to 1.3 for consistency with other products and introduced the installation factor to be set at 1.5 for typical installation practices. Lower values could be used if strict installation controls were implemented. AASHTO did not adopt the reduced load factor but did adopt the recommended installation factor. Thus there is no need to increase the installation factor above the value of 1.0 when the load factor is taken as 1.95.

The AASHTO LRFD load factor for water is 1.0. For groundwater permanently at or above ground level, the buoyant weight of the backfill should be used for γ_s. For intermediate groundwater levels, the wet unit weight or buoyant unit weight should be used as appropriate. For any condition, the groundwater level used in the design should be selected to give an appropriately conservative estimate of thrust.

If the composite-constrained soil modulus considering trench wall stiffness is used for deflection, it should also be used in the thrust calculation.

The live load factor F_1 was introduced for metal conduits. The live load factor F_2 accounts for reduction in thrust due to the low hoop stiffness ratio.

Local buckling – The limiting criteria for corrugated PE and PP conduits, which have low cross-sectional area and low long-term modulus of elasticity, is usually local buckling. The theory for buckling of plates was developed by Bryan (1891) and adapted to light gauge metal sections by Winter (1946). The basis of Winter's work is that the supported edges of a plate element have additional capacity after the center of the plate buckles (Figure 7.38a) even though the central portion of the plate becomes ineffective, as shown in Figure 7.38b. The ultimate condition is reached when a compression failure occurs in the braced corner sections. The width-to-thickness ratio (w/t) of the unbraced element and the magnitude of the applied stress are the key parameters in determining the effective width. Element width, w, is the clear distance between the supports of the element (Figure 7.38b). After analysis for local buckling, the net effective area, A_{eff}, is used in the thrust equation.

McGrath and Sagan (2000) adapted Winter's approach to the design of corrugated plastic profile wall conduits. A typical corrugated profile wall is shown in Figure 7.39, along with an idealized profile for analysis of local buckling capacity. Using Winter's theory, each element could be considered a plate and evaluated for local buckling under compression stresses.

The effective area of a conduit profile geometry is conducted by subtracting the ineffective area of each element from the total area of the cross-section as follows:

$$A_{eff} = \frac{A_g - \Sigma(w - b_e)t}{\omega}, \tag{7.50}$$

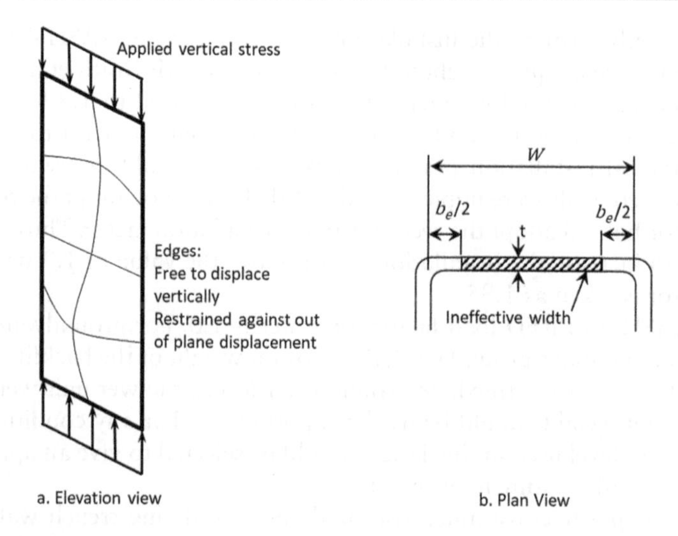

Figure 7.38 Buckling of thin elements with edge support.

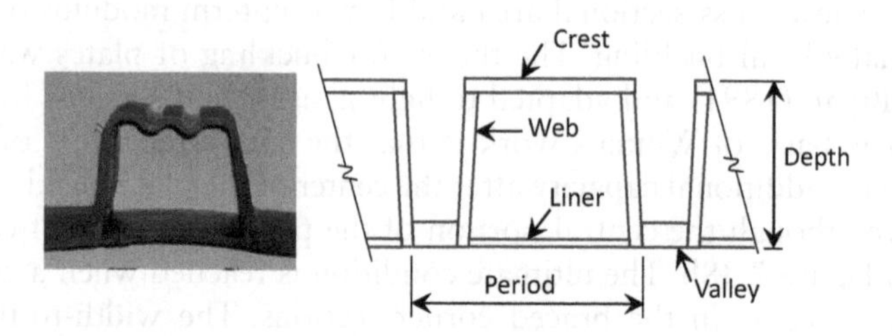

Figure 7.39 Typical idealization of profile shape and corrugation terminology.

in which for each element

$$b_e = \rho w, \tag{7.51}$$

$$\rho = \frac{\left(1 - \frac{0.2}{\lambda}\right)}{\lambda}, \tag{7.52}$$

$$\lambda = \left(\frac{w}{t}\right)\sqrt{\frac{\varepsilon_{yc}}{k}} \geq 0.673, \tag{7.53}$$

where

A_{eff} = Effective area of corrugated profile per unit length, in.^{2}/in., mm^2/mm
A_g = Total area of corrugated profile per unit length, in.^{2}/in., mm^2/mm
w = Clear width of element between supporting edges, in., mm
b_e = Effective width of element, in., mm
t = Thickness of element, in., mm
ω = Period of corrugation, in., m
ρ = Effective width factor
λ = Slenderness factor
ε_{yc} = Factored compression strain limit
k = Plate buckling coefficient, k = 4 for supported elements (all elements in Figure 7.38), k = 0.43 for unsupported elements, such as free-standing ribs.

For profiles such as those shown in Figures 7.39 and 7.40, the liner is generally so thin as to be mostly ineffective in resisting compression stress. The primary function of the liner is to provide a smoother inner surface to improve hydraulic flow. The valley, which is a composite of the liner and corrugation, is typically fully effective, while the crest and webs are partially effective.

Some profiles are not easily idealized, such as Figure 7.40a, with an arced crest. Curved sections are resistant to buckling, but with experimental evidence (see the following discussion on the stub compression test), a suitable idealization could be developed. Figure 7.40b suggests one option, with the top element set at the centroid of the arc portion of the conduit. The calculation can be calibrated against testing with the stub compression test.

A second method of evaluating the compression strength of a profile (i.e., its ability to resist local buckling) is the "stub compression test," AASHTO Standard T341 *Standard Method of Test for Determination of Compression*

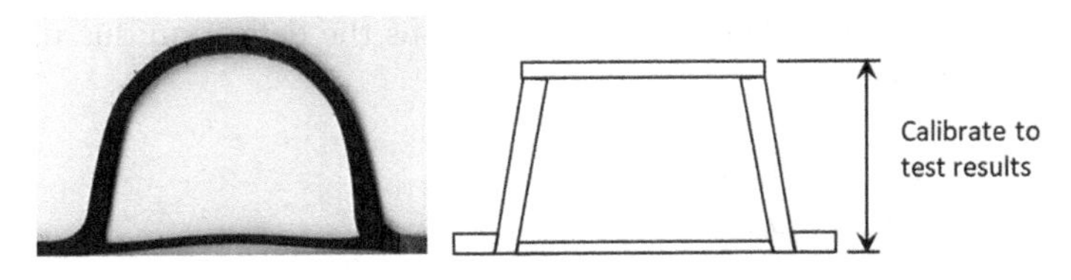

Figure 7.40 Idealization of nonrectangular profile.

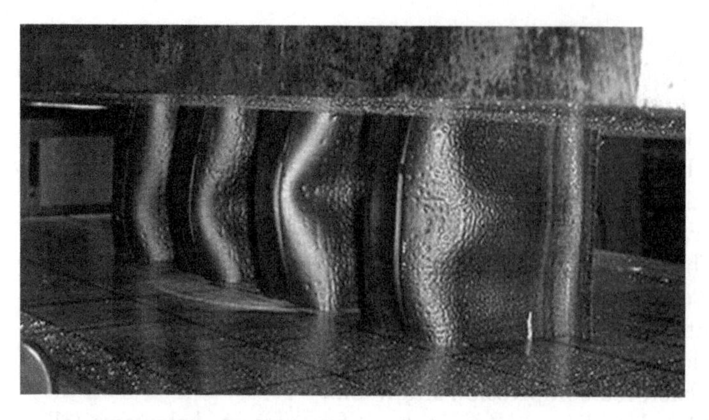

Figure 7.41 Local buckling in corrugated conduit wall sections.

Capacity for Profile Wall Plastic Pipe by Stub Compression Testing (Figure 7.41). Features of the test include the following:

- Specimens are taken from conduit walls, generally at least three corrugations long. The circumferential portion must be short to minimize bending stresses, generally about 1.5 times the depth of the profile.
- Preparation of the test specimens requires considerable care to machine the bearing ends to a flat plane. Uneven specimen ends will result in high localized stresses and premature buckling of some elements.
- The test is conducted by compressing the specimen between two fixed plates at a low crosshead speed, increasing the load until a maximum is reached. The crosshead displacement during loading divided by the specimen height prior to testing is the average strain at failure. Changing crosshead speed does not change the test result as buckling is a geometric phenomenon. That is, the effective modulus at the time of maximum load is not a factor in the test result.

Flexural strain – Flexural strains, ε_f, in thermoplastic conduits are calculated based on deflection level and a shape factor, D_f, that accounts for nonelliptical deformation. The deflection of interest is only due to bending, Δ_f, which is the allowable deflection, Δy, minus the deflection due to hoop compression, ε_{sc} D_m.

$$\varepsilon_r = \gamma_{EV} D_f \left(\frac{c}{R_m}\right)\left(\frac{\Delta_f}{D_m}\right), \tag{7.54}$$

where

γ_{EV} = load factor for vertical earth load
D_f = shape factor, Table 7.17

c = larger of the distance from the centroid of conduit wall to the inner or outer extreme fiber, in., mm
R_m = radius to centroid of conduit wall, in., mm
Δ_f = Vertical deflection due to bending
= $\Delta_y - \varepsilon_{sc} D_m$
Δ_y = allowable vertical deflection, in., mm
ε_{sc} = compression strain in conduit wall due to hoop thrust, in./in., mm/mm (see Section 7.7.2)
D_m = diameter to centroid of conduit wall, in., mm

The shape factor is an empirical term that defines the distortion of the conduit shape relative to deformation in the parallel plate test (Equation 3.11, repeated here).

$$\varepsilon = 4.27 \frac{t}{D_m} \frac{\Delta Y}{D_m} \tag{7.55}$$

If $D_f = 4.27$, the bending strain would be the same as in the parallel plate test. It is common to think of the parallel plate test as a severe condition representative of flat bedding, but this is not the case. In the parallel plate test, the top and bottom of the conduit deform identically, while in the case of a conduit installed on flat hard ground, the top of the conduit, which is surrounded by backfill will likely deform into an elliptical shape while the invert flattens against the hard bedding. Thus, most of the conduit deflection will occur in the lower half of the conduit, and the shape factor, D_f, could be doubled since it is based on the total change in diameter. Suggested values for D_f are presented in Table 7.17.

Table 7.17 is taken from AWWA Manual M45 (AWWA, 2014) for fiberglass conduit and shows

- decreasing values as conduit stiffness increases because low-stiffness conduits are more readily deformed during placing and backfilling operations,
- increasing shape factor with backfill of smaller particle size (and increased fines content), and
- increased shape factor with increased compaction levels.

Conduit distortion can occur due to the bedding condition, the way soil is placed in the trench, for example, dumped from ground level or ladled in from an excavator bucket, the type of compaction equipment, how close it is operated to the conduit and how many passes are made. There are likely other construction practices that could affect distortion as well.

Combined strain – Local buckling can also occur due to deflection and resulting flexural stresses. However, testing and computer modeling have

Table 7.17 Shape factor, D_f, based on conduit stiffness, backfill type, and compaction level

Conduit Stiffness EI / 0.149 R_m^3, psi(kPa)	*Conduit Zone Embedment Material and Compaction Level*			
	Gravel		*Sand*	
	Dumped to Slight	*Moderate to High*	*Dumped to Slight*	*Moderate to High*
9 (62)	5.5	7.0	6.0	8.0
18 (125)	4.5	5.5	5.0	6.5
36 (250)	3.8	4.5	4.0	5.5
72 (500)	3.3	3.8	3.5	4.5

AWWA, 2014, reprinted with permission. M45 Fiberglass Pipe Design, 3rd edition, American Waterworks Assn.

shown that the limiting strain is greater than due to thrust alone because flexural strains in the web elements are low as the neutral axis is near the middle of the web. This increases the overall stability of the profile. Thus, AASHTO LRFD allows a 50% increase in the ultimate compression strain for the combined condition.

Buckling – Global buckling is evaluated with the same equation used for deep corrugated metal conduits except that it is expressed in terms of strain, Moore and Selig (1990):

$$\varepsilon_{bck} = 1.2\phi_b C_n \left(E_p I_p\right)^{\frac{1}{3}} \left(\phi_s M_s k_b\right)^{\frac{2}{3}} R_h, \tag{7.56}$$

where

R_b = nominal axial force in conduit wall to cause general buckling, lb/in., kN/m
ϕ_b = resistance factor for general buckling
C_n = 0.55, scalar calibration factor to account for some nonlinear effects (Moore, 1989)
E_p = modulus of elasticity of conduit wall material, ksi, KPa
I_p = moment of inertia of conduit wall, including stiffeners if used, in.^{4}/in., m^4/m
ϕ_s = resistance factor for soil
M_s = constrained modulus of embedment based on free field vertical soil stress at a depth halfway between crown and springline, ksi, KPa
k_b = soil stiffness correction factor
= $(1 - 2\nu)/(1-\nu^2)$
ν = Poisson's ratio of soil
$R_h = 11.4 / (11 + S / H)$, correction factor for backfill geometry

S = conduit maximum span, ft, m
H = depth of fill over top of conduit, ft, m

The complete theory developed by Moore and Selig (1990) presents alternate values of R_h for nonuniform embedment conditions.

7.7.3 Fiberglass conduits

Fiberglass conduit design is almost identical to thermoplastic conduit design except that the compression strains are small and are not considered in computing conduit wall strains. The design of fiberglass conduits is covered in detail in *AWWA Manual of Practice M45 Fiberglass Pipe Design, 3rd Edition* (AWWA, 2014).

REFERENCES

AASHTO (2020) *AASHTO LRFD Bridge Design Specifications*, 9th Edition, AASHTO, Washington, DC.

AASHTO (2023) *AASHTO LRFD Construction Specifications*, 4th Edition, AASHTO, Washington, DC.

ACPA (1998) *Concrete Pipe Design Handbook*, American Concrete Pipe Association, Irving, Texas.

ACPA/CCPA (2023) PIPEAC, American Concrete Pipe Association and Canadian Concrete Pipe Association, http://pipe.concretepipe.org/

ASCE/CI (2000) *ASCE/CI 28-00 Standard Practice for Direct Design of Precast Concrete Box Sections for Jacking in Trenchless Construction*, ASCE, Reston, VA.

ASCE/CI (2017) *ASCE/CI 27-17 Standard Practice for Direct Design of Precast Concrete Pipe for Jacking in Trenchless Construction*, ASCE, Reston, VA.

AWWA (2014) *AWWA Manual M45 Fiberglass Pipe Design*, 3rd Edition (errata), AWWA.

Bryan, G.H. (1891) On the Stability of a Plane Plate Under Thrusts in Its Own Plane, with Applications to the "Buckling" of the Sides of a Ship, *Proceedings of the London Mathematical Society*, Vol. 22.

Burns, J.Q., and Richard, R.M. (1964) Attenuation of Stresses for Buried Circular Cylinders, *Proceedings of the Symposium of Soil Structure Interaction*, University of Arizona, Tucson, AZ, pp. 378–392.

CANDE (2022) Culvert ANalysis and DEsign, https://www.candeforculverts.com/home.html

Chambers, R.E., McGrath, T.J., and Heger, F.J. (1980) *Plastic Pipe for Subsurface Drainage of Transportation Facilities, NCHRP Report 225*, National Academy of Sciences, Washington, DC.

Duncan, J.M., Seed, R.B., and Drawsky, R.H. (1985) *Design of Corrugated Metal Box Culverts (Transportation Research Record 1008)*, Transportation Research Board, Washington, DC.

Eriksson (2021a) *Eriksson Culvert*, Eriksson Software Inc., Temple Terrace, FL.

Eriksson (2021b) *Eriksson Pipe*, Eriksson Software Inc., Temple Terrace, FL.

Galambos (1981) Load and Resistance Factor Design, *Engineering Journal of the American Institute of Steel Construction*, American Institute of Steel Construction. Chicago, IL.

Heger, F.J., and McGrath, T.J. (1982) *Design Method for Reinforced Concrete Pipe and Box Sections*, Report to the Technical Committee of the American Concrete Pipe Association, Rev. 1982.

Heger, F.J., and McGrath, T.J. (1982a) Shear Strength of Pipe, Box-Sections and Other One-Way Flexural Members, *Journal of The American Concrete Institute*, Vol. 79, No. 6.

Heger, F.J., and McGrath, T.J. (1983) Radial Tension Strength of Pipe and Other Curved Flexural Members, *Journal of the American Concrete Institute*, Vol. 80, No. 1.

Heger, F.J., and McGrath, T.J. (1984) Crack Width Control in Pipe and Box Sections, *Journal of the American Concrete Institute*, Vol. 81, No. 2.

Katona, M.G., and Akl, A.Y. (1987) Structural Design of Buried Culverts and Slotted Joints, *Journal of Structural Engineering*, Vol. 113, No 1, pp. 44–60.

Liu, Y., Moore, I.D., and Hoult, N.A. (2024) New Design Equations for Corrugated Steel Culverts Responding to Live Loads, *Canadian Geotechnical Journal*, Vol. 61, No. 1, pp. 1–15.

Marston, A. (1930) *The Theory of External Loads on Buried Conduits in the Light of the Latest Experiments, Bulletin 96*, Iowa State College, Ames, IA.

McGrath, T.J., and Chambers, R.E. (1981) Field Performance of Buried Plastic Pipe, *Underground Plastic Pipe, Proceedings of the Conference*, ASCE.

McGrath, T.J., Liepins, A.A., and Beaver, J.L. (2005) Live Load Distribution Widths for Reinforced Concrete Box Sections, *Transportation Research Record: Journal of the Transportation* Research *Board, CD-11-S*, pp. 99–108.

McGrath, T.J., Moore, I.D., Selig, E.T., Webb, M.C., and Taleb, B. (2002) *NCHRP Report 473 Recommended Specifications for Large-Span Culverts*, National Academy of Sciences, Washington, DC.

McGrath, T.J., Moore, I.D., and Hsuan, G.Y., (2009) *NCHRP Report 631 Updated Test and Design Methods for Thermoplastic Drainage Pipe*, National Academy of Sciences, Washington, DC

McGrath, T.J., and Sagan, V.E. (2000) *NCHRP Report 438 Recommended LRFD Specifications for Plastic Pipe and Culverts*, National Academy of Sciences, Washington, DC.

Moore, I.D. (1989) Elastic Buckling of Buried Flexible Tubes – A Review of Theory and Experiment, *Journal of Geotechnical Engineering*, American Society of Civil Engineers, Reston, VA, Vol. 115, No. 3, pp. 340–358.

Moore, I.D., Hoult, N.A., and MacDougal, K. (2014) *Establishment of Appropriate Guidelines for Use of the Direct and Indirect Design Methods for Reinforced Concrete Pipe*, Prepared for the AASHTO Standing Committee on Highways, Queens University, Kingston, Ontario.

Moore, I.D., and Selig, E.T. (1990). Use of Continuum Buckling Theory for Evaluation of Buried Plastic Pipe Stability. *Buried Plastic Pipe Technology, ASTM STP 1093*, George S. Buczala and Michael J. Cassady, Eds., American Society for Testing and Materials, Philadelphia, PA, pp. 344–359.

Petersen, D.L., Nelson. C.R., Li, G., McGrath, T.J., and Kitane, Y. (2010) *Recommended Design Specifications for Live Load Distribution to Buried Structures, NCHRP Report 647*, National Academy of Sciences, Washington, DC.

Vecchio, F.J., and Collins, M.P. (1986) Modified Compression Field Theory for Reinforced Concrete Elements Subjected to Shear, *ACI Journal*, Vol. 83, No. 2, pp. 19–231, American Concrete Institute, Chicago, IL.

Winter, G. (1946) Strength of Thin Steel Compression Flanges, *ASCE Transactions Vol. 112* pp. 339–387. American Society of Civil Engineers, Reston, VA.

Chapter 8

Installation and construction monitoring

Once a project has been designed, specifications written, and a contractor selected, the next step is installation, where the design is implemented. Prior to executing a contract for construction, the owner and the owner's representative should have made decisions on responsibilities for activities required during and after construction:

- Verification that the conduit and backfill meet the specifications
- The methods used for monitoring during construction
- Postconstruction inspection
- The parties responsible for construction monitoring and postconstruction inspection

During construction, the contractor must complete the following:

- Purchase and install the specified materials in accordance with the specifications.
- Provide training and supervision for crews executing the work.
- Provide a safe working environment.

The project specifications define the project being built and the responsibilities of the parties involved. While the contractor and owner may be the only signatories to the construction contract, the owner's representative, most often the engineer, is typically referenced and assigned responsibility as well. Subcontractors involved in conduit handling, placement, and backfilling should also be involved in preconstruction activities.

Sometimes questions arise during construction on various details of the project. A responsible approach when this happens is to first determine if the situation is addressed in the contract. If disputes arise during or after construction, it is the written record, i.e., the contract and related correspondence, that will be decisive in reaching a settlement. For this reason, constant reference to the contract documents is important and agreements or changes made during the construction period should be documented in writing to

DOI: 10.1201/9780429162619-8

establish the nature of the change and the agreement of the parties to any deviations from the contract.

This chapter will focus on how construction practices affect the performance of the structural system (and the impact of soil-conduit interaction). Some important construction activities are addressed, but further details may be found in the many existing installation guides (ASTM standards, AASHTO specifications, and books such as Howard, 2015). For example, the necessity of dealing with groundwater problems is called out, but details of operating a dewatering system during installation are not. Also, trenching operations must be conducted in a way that provides safety for workers in the trench and anywhere on the site; again, the details of how that is accomplished are not addressed except that discussion is included on how it might affect the conduit performance. This chapter is intended to provide guidance on conduit installation and does not purport to address all issues with sufficient detail to constitute a full specification. The AASHTO LRFD Construction Specifications (AASHTO, 2017), Articles 26, 27, and 30 address installation of metal, concrete, and plastic pipes, respectively.

8.1 SITE CONDITIONS AND PERSONNEL SAFETY

The contract should alert the contractor to known site conditions. This may include providing soil borings, but it may be appropriate to identify other items the contractor should consider before bidding on the project:

- Control of groundwater may be required. If the engineer determines that the system will require well points outside of the trench to keep it dry enough for pipe-laying operations, then methods of dewatering from within an open trench should be prohibited.
- If the site requires overexcavation at some locations to replace soft or wet soil, the contract should identify who makes that decision and should provide the expected details of the remediation, such as the nature of the replacement material and if a geotextile fabric will be required. With these details, the contractor can offer a unit price for such work within the bid.

There may be other such issues that arise during construction. Identifying such items in advance can save time and money.

The contractor has control of the site where the work is taking place and should always be responsible for the safety of employees and others who access the site. A typical contract makes this clear. It is particularly important to avoid contract clauses that shift site safety responsibility away from the contractor unless there is an extenuating circumstance.

8.2 BACKFILL AND TRENCH TERMINOLOGY

Terminology for trench zones was presented in Figure 2.2. This figure is presented again here for ease of reference.

Generally, flexible conduits use a single soil for the sidefill and topfill. Some flexible conduits use a "split installation" where the sidefill extends to a height to encompass the region where the conduit wall relies on soil support and a lower quality fill for the topfill. Such installations require care that the two materials are compatible, particularly concerning migration of fines from the finer material to the coarser material. See Section 9.2 for a project where this was not the case and substantial failures occurred.

Concrete uses different backfill zones and terminology than flexible conduits.

- The haunch zone extends beyond the area below the pipe springline.
- The "lower side" is introduced for embankment installations to identify an area where compaction is required to prevent arching of load on to the sidefill and conduit.
- Backfill above the springline is not highly controlled.

8.2.1 Embedment zone soil groups

There are many ways to specify backfill materials. Chapter 4 introduced several of the systems for classifying backfill materials and their groupings based on stiffness and compactability. Some classifications are broad, such as SW per ASTM D2487 *Standard Practice for Classification of Soils for Engineering Purposes (Unified Soil Classification System)*, while others, such as AASHTO M43 *Standard Specification for Sizes of Aggregate for Road and Bridge Construction* use very specific gradations. For example, as shown in Figure 8.2, ASTM D2487 soil group SW has particle sizes up to 3 in. (75 mm) but otherwise only restricts relative content of the particles passing the No. 200 sieve (0.075 mm) to less than 5%, with requirements to have a higher percentage of sand than gravel, and to meet the requirements for uniformity, C_u, and curvature, C_c, to establish that it qualifies as well graded. AASHTO M43 No. 57 stone has a limited gradation between the lines "No. 57 max" and "No. 57 min" in Figure 8.2. No. 57 stone is sometimes specified as a conduit backfill material as it provides good stiffness with little or no compactive effort.

Broad classifications are desirable because they are less restrictive and allow more possible materials for a contractor to select from, including both natural and processed materials. Most conduit installation standards use broad classifications. Specific gradations are more definitive but also more expensive, as they are processed to specific gradations to obtain benefits, such as high compactability and ease of placement. Designers should

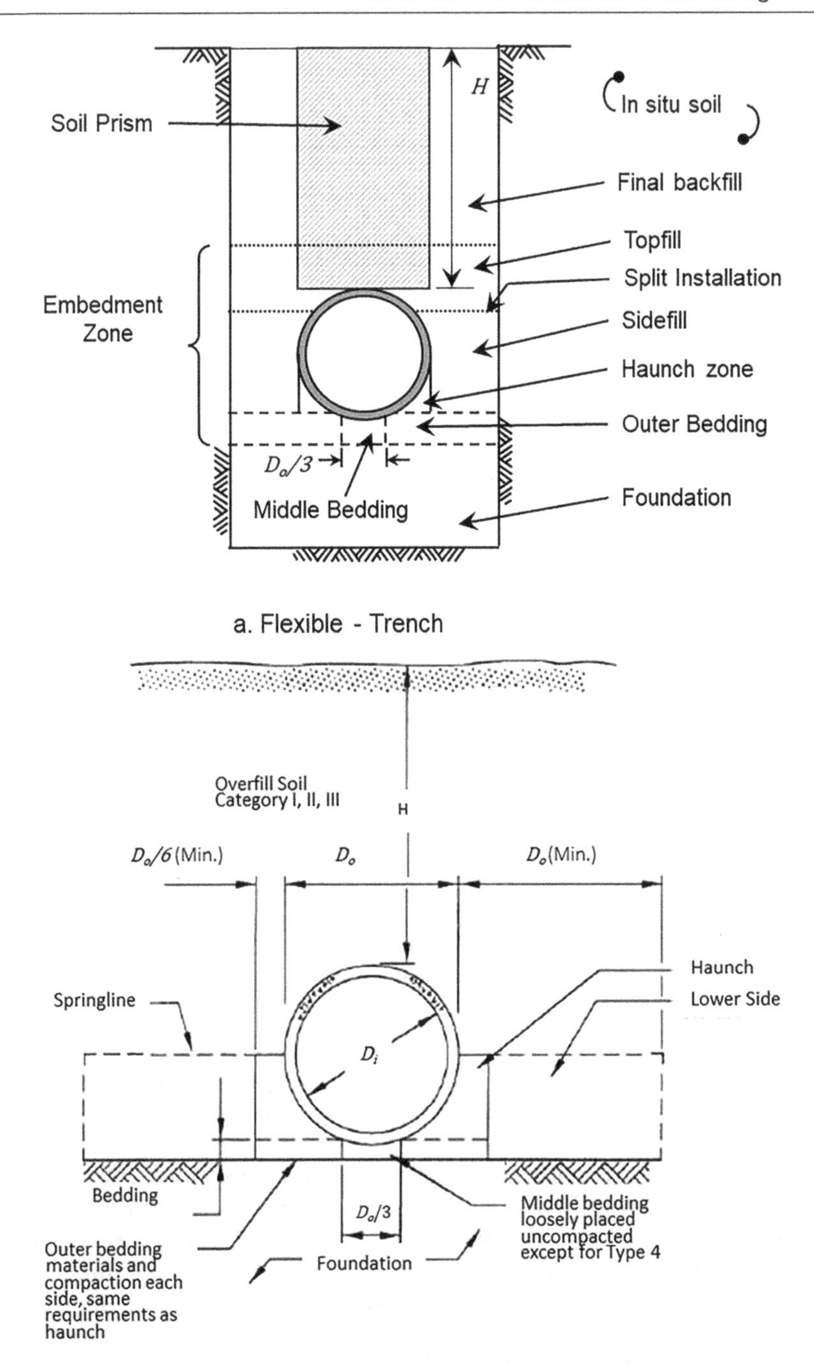

Figure 8.1 Terminology for trench and embankment installations.

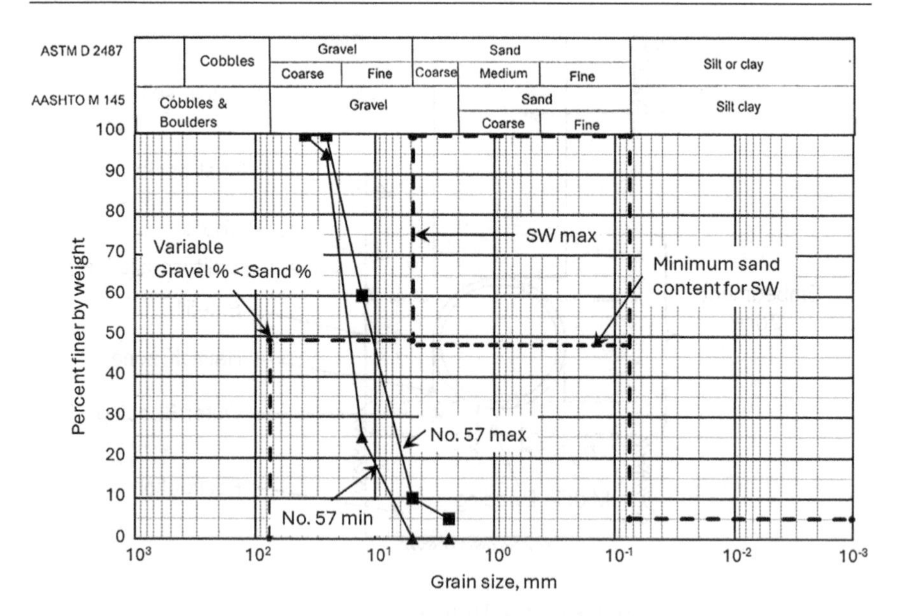

Figure 8.2 Gradation limits for SW soil and No. 57 stone.

determine if limited gradations, such as No. 57 stone, will provide sufficient benefit to justify the cost.

Smaller diameter conduits should have a smaller maximum particle size, as thinner walls might be susceptible to damage. ASTM D2321 *Standard Practice for Installation of Thermoplastic Pipe for Sewers and Other Gravity Flow Applications* sets maximum particle size as 1½ in. (37 mm), but an even smaller size may be appropriate for smaller conduits. For conduits with corrugated exteriors, it is recommended to have the maximum particle size limited to allow filling of the space between corrugations.

Embedment zone soil selection is based primarily on compactability and achievable soil stiffness. Estimating the cost of backfilling a trench should consider the cost of both the materials and the labor and equipment to place and compact it. Less expensive materials will generally require more compactive energy and greater effort to work into the haunch zone (see Section 4.3.5), which in turn may require increased inspection, all of which contribute to the total cost. Soil groups based on compactability, and soil stiffness were presented in Table 4.9, Section 4.3.4, which is reproduced here as Table 8.1 for convenience. The soil groups in each column vary because of the different classification systems. The approach here is to provide groups that are easily specified rather than introduce adjustments to make all soil groups identical. The contract documents should require the contractor to submit specific gradations for approval to allow the owner/engineer to reject unusual gradations that may prove unsuitable as backfill.

The AASHTO soil groups were developed to classify soils according to their suitability as roadway subgrade, but that use has been extended to include use as conduit embedment. With this focus, the classifications do not align with the ASTM groups, which are classified more strictly based on behavior. Table 8.1 allows these differences because each group has sufficiently similar compatibility and stiffness and can be used with the same M_s or E' values with sufficient accuracy for design. For ease of reference in the following discussion, soil groups are referred to by the D2321 group names (Class I, II, etc.), and material passing the No. 200 sieve (0.075 mm) is called fines.

Class I – Class I materials are open-graded, coarse-grained materials and can sometimes reach a design stiffness value when dumped. Angular materials must be worked into the haunch zone. Rounded materials will more readily flow into the haunch zone but still require some tamping to achieve the desired support. If the surrounding materials are at risk of migration into the voids in Class I material, a geotextile filter fabric is recommended. If used as a foundation material to replace unacceptable in situ trench bottom soil, a geotextile filter fabric may be required to prevent mixing of the open-graded material with the in situ soil. When using the M_s table for soil stiffness, it is common to use the value for Sn95 soil if the material is dumped and the Sn100 material if it is compacted. Controlling the moisture content of Class I materials is not necessary.

Class II – Class II materials are coarse-grained soils per ASTM or granular soils per AASHTO. The ASTM soils have fines limited to 12% or less. AASHTO A-1 soils can have up to 25% fines, but the amount passing a No. 40 sieve (0.425 mm) is restricted to 50%, assuring a significant gravel content. AASHTO A-3 soils are fine sand requiring a minimum of 51% passing the No. 40 sieve and allowing up to 10% passing the No. 200 (0.075 mm) sieve. The suggested limit on material passing the No. 100 (0.15 mm) sieve limits the total fine sand plus silt content to an acceptable level to improve compactibility and limit the risk of the fine material migrating or flowing during or after construction. Class II soils achieve good stiffness with moderate compactive effort. Required control of moisture is generally minimal, but this should be verified at the beginning of construction.

Class III – Class III soils have increased fines content, require more attention to control water content, and achieve lower soil stiffness than Class I or II soils. Table 8.1 allows GM, GC, SM, and SC soils, which have up to 49% fines, while ML and CL are acceptable if the fines content is restricted to 70% and the liquid limit (LL) to less than 50. AASHTO soils A-2-4 and A-2-5 have fines limited to 35% and are differentiated based on the LL and plasticity index (PI). Class III soil must have the water content controlled to near the optimum value to achieve specified

Table 8.1 Soil groups for conduit embedment (previously presented as Table 4.9)

ASTM Soil Groups[A]	*AASHTO Soil Group*[B]	*Group Names*[C] *(Concrete/Thermoplastic/Fiberglass)*	M_s *Group*
Crushed rock[D]: 100% passing 1-1/2 in. (37 mm) sieve, ≤ 15% passing No. 4 (4.75 mm) sieve, ≤ 25% passing 3/8 in. (9.5 mm) sieve and ≤ 12% passing No. 200 (0.075 mm) sieve	...	– Class I SC1	Sn[G]
Clean coarse-grained soils[E]: SW, SP, GW, GP, or any soil beginning with one of these symbols with ≤ 12% passing No. 200 (0.075 mm) sieve[C]	A1, A3[F]	Category I Class II SC2	Sn
Coarse-grained soils with fines: GM, GC, SM, SC, or any soil beginning with one of these symbols, containing > 12% passing No. 200 (0.075 mm) sieveSandy or gravelly fine-grained soils:ML, CL, or any soil beginning with this symbol, with ≥ 30% retained on the No. 200 (0.075 mm) sieve	A-2-4, A-2-5	Category II Class III SC3	Si
Fine-grained soils: CL or ML with < 30% retained on No. 200 (0.075 mm) sieve or any soil beginning with one of these symbols	A-4, A-2-6, A-2-7	Category III Class IV SC4	Cl
MH, CH, OL, OH, PT	A5, A6, A7	– Class V SC5 Not for use as embedment	

[A] Based on ASTM D2487 with some additions and modifications to improve suitability as embedment materials.
[B] Based on AASHTO M145.
[C] Group names for concrete, thermoplastic, and fiberglass conduits are from AASHTO LRFD Article 12, ASTM D2321, AWWA Manual M45, respectively. Metal conduit design practice does not use specific group names.
[D] Class I materials are labeled crushed rock, but rounded materials such as pea gravel can also provide high stiffness with minimal compaction.
[E] Materials such as broken coral, shells, and recycled concrete, with ≤ 12% passing a No. 200 (0.75 mm) sieve, are considered Class II materials. These materials should only be used when approved by the engineer, as there may be additional considerations concerning long-term suitability.
[F] A-3 soils are restricted to a maximum of 50%, passing a No. 100 (0.15 mm) sieve. Restrictions should be considered on the portion passing the No. 100 sieve to avoid a material overly sensitive to moisture.
[G] M_s for crushed rock is often taken as Sn values at 95% and 100% maximum Standard Proctor density for the dumped and compacted conditions. Howard (2015) suggests specific E' values for crushed rock.

compaction levels. They require more compactive effort to reach specified unit weights and achieve less stiffness than Class II soils at the same percent of maximum standard Proctor density.

Class IV – Class IV soils in general have high fines content, with increased limits on LL and PI. A-2-6 and A-4 soil can have up to 64% fines, but the LL and the PI are limited. These materials have more limited applications, as they require high compaction levels to achieve moderate stiffness levels and must have the water content controlled during placement and compaction. Class IV soils are difficult to work into and compact in the haunch zone. AASHTO, which focuses on roadways, does not consider these materials suitable for conduit embedment.

Class V – Class V materials should not be used as conduit embedment.

8.2.2 Migration

Whenever two different soils are placed next to each other, the relative gradations must be assessed to determine the risk of mixing over time, which will reduce the total soil volume and thus the soil support to the conduit. Section 9.2 presents one case where migration contributed to significant conduit failures.

The migration process is generally driven by movements of groundwater, either infiltration of rainwater moving downward through the soil and perhaps through conduit joints if they are not watertight or longitudinal groundwater movement along the conduit to the outlet. Long pipelines with open-graded embedment can be susceptible to groundwater movement of finer materials if the water flows along the pipeline through the embedment to some exit point. This likelihood can be reduced or eliminated if the open-graded backfill is periodically interrupted with a short length of a material with low hydraulic conductivity.

One set of criteria to assess the relative gradation of two soil types is

- $D_{15}/d_{85} < 5$, where D_{15} is the sieve-opening size passing 15% by weight of the coarser material, and d_{85} is the sieve-opening size passing 85% by weight of the finer material, and
- $D_{50}/d_{50} < 25$, where D_{50} is the sieve-opening size passing 50% by weight of the coarser material, and d_{50} is the sieve-opening size passing 50% by weight of the finer material. This criterion need not apply if the coarser material is well graded as defined in ASTM D2487.

In addition to using these criteria to determine if separation is required between embedment and in situ soils, they can also be used to specify an intermediate soil between the embedment and in situ soils. This layered approach can replace the use of a geotextile, but it must be assessed for ease of construction. Straight-line boundaries between soil groups can be presented on contract drawings but are not always so easy to construct.

Filter fabrics are preferred for maintaining separation between incompatible materials as the use of filters requires controlled gradations and can be difficult to place.

8.3 EXCAVATION

Most conduits are installed in trenches, but most of the provisions for foundation, bedding, and backfill also apply to an embankment installation. The key point for embankment installations is that the sidefill must extend far enough out from the conduit springlines to encompass the soil-conduit interaction zone, or the embankment material at the sides of the embedment material will need to have some minimum stiffness, as indicated in Figure 8.1b for concrete pipe installations.

Excavation must be carried out in accordance with all applicable laws and guidelines for providing a work site that is always safe for personnel. This may require sloped trench walls or trench wall supports. This includes workers in the trench but also includes support workers not in the trench. Heavy equipment operations, the lifting and placing of conduit sections, and other activities all pose significant risks to anyone on the work site.

8.3.1 Trench width

When conduits are installed in trenches, contract provisions should require a minimum trench width that assures the ability of workers to place and join conduit segments, backfill below the springline and work it into the haunch zone, and operate compaction equipment at the springline. This width will increase as the conduit diameter increases.

Specifications should also provide a maximum trench width. If the load on the conduit was based on a certain trench width, a wide trench could increase the load. The specified trench width should likely be smaller than assumed in the design to minimize situations where the load will increase to greater than the design assumption due to a wide trench. If sloped trench walls are used, the elevation at the top of the conduit is the location to determine trench width. The contractor should be alerted in the contract if a trench width was assumed in design and of any consequences if the trench width is exceeded.

8.3.2 Trench wall supports

Trench wall supports must not interfere with the conduit or the embedment supporting the conduit. Balancing the need for worker safety and proper conduit installation can be challenging. The primary concerns are presented here. More details and options are presented in texts devoted to pipeline installation such as Howard (2015).

- Trench boxes – It is preferable to keep the bottom of trench boxes at least above the top of the conduit springline to minimize disrupting the embedment when moving the trench box. When it is necessary for the trench box to extend below the top of the pipe, it must be raised during backfilling or moved forward without disrupting the trench or the embedment, and the gaps from the trench box sides must be filled with compacted embedment soil.
- Sheeting should have tight joints to prevent water from flowing into the trench. If the trench passes under an existing utility or conduit, steps must be taken to brace the excavation and limit water intrusion under the passing conduit where sheeting cannot be driven.

When sheeting is installed below the top of the pipe, it should be left in place to prevent disruption of the embedment. If sheeting is removed, it must be verified that the conduit embedment is not disrupted.

8.4 TRENCH PREPARATION

8.4.1 Foundation

If the bottom of the trench is unsuitable for pipe support or inadvertently overexcavated, it will need to be remediated to a sound condition. Unsuitable conditions could include the presence of organic soil, soft soil, or excess water. The means and responsibility for assessing these conditions should be addressed in the contract.

Remediation of the trench bottom should provide a stable, dry foundation on which to place the bedding, conduit, haunching, and sidefill. Remediation of the trench bottom should extend to the full width of the trench. This is required because the foundation will support both the conduit and the embedment material. If the embedment material under the sidefill is not properly supported, it can settle and increase the load on the conduit (raising the vertical arching factor [*VAF*]). The concern for vertical support of the embedment material is also applicable to embankment installations. See Section 9.5 for a case study where such support was not provided.

If soil in the trench bottom is suitable but loose and dry, it may be feasible to compact it to a higher unit weight. The stiffness after compaction should be consistent with that specified for the embedment material.

If the soil in the trench bottom is unsuitable or too wet, it will need to be removed to sound soil and replaced. As noted earlier, the replacement material should be compacted to provide a stiffness (M_s or E') similar to that specified for the sidefill. If the unsuitable soil is dry or not saturated, it can be replaced with the material specified for the bedding. If the unsuitable condition is wet, an open-graded material may be required to control the condition. There are several items to consider when placing an open-graded foundation material:

- In some conditions, working the material into the soft material will stabilize it sufficiently to support the bedding, conduit, and embedment. This avoids the need for geotextiles and the risk of migration into the foundation material.
- A geotextile filter fabric can be placed under and around the foundation material to prevent the migration of fines.
- In some conditions, a reinforcing geotextile fabric can be used to improve support of the foundation material.
- If the condition is a steady flow of water, placing a pump in a sump hole may be required to control the water until backfilling is completed above the groundwater level. Removing water may also be facilitated by placing a perforated pipe within the open-graded material. Any pumping operations must be designed to minimize removal of fines.
- In some situations, it may be necessary to dewater outside the trench with well points.

8.4.2 Bedding

Bedding is the surface a conduit is placed on. The bedding provides a cushion to protect a conduit from hard native materials in the trench bottom. In some natural soil deposits, a conduit can be laid directly on the trench bottom, but this is generally not recommended, as native materials can vary unpredictably along the route of the conduit. The stiffness of the natural soil may vary and may have hidden rocks or hard points just below the surface that could impinge on the pipe as the earth load is placed over the top. Recommended bedding depth typically varies from 3 to 6 in., depending on the type of conduit and, for some products, the conduit diameter.

Most standards call for increasing the bedding thickness if the trench bottom is rock or other unyielding material. Hard trench bottoms, especially rock, can be uneven, and extra thickness minimizes the possibility that rock projecting up from the trench bottom will impinge on the conduit.

If the conduits have expanded bells (outside diameter of the bell is greater than the barrel), then small depressions must be created in the bedding to permit the barrel to be uniformly supported.

8.4.2.1 Round conduits

Bedding for round conduits is particularly important, as it can spread the soil pressure distribution on the bottom of the pipe over a greater width. Hard bedding results in a concentrated load at the invert of a conduit, which increases local stresses in the pipe wall, while soft or shaped bedding distributes the pressure over a wider portion of the pipe (Figure 8.3).

The bedding can be placed and shaped prior to placing the conduit, but that is time-consuming, and it is difficult to achieve good conformance of the bedding to the conduit shape. Leaving the middle bedding (Figure 8.1) uncompacted can achieve the same benefit with much less effort. The standard beddings for concrete conduits in the AASHTO LRFD Bridge Construction Specifications (AASHTO 2017) call for the middle bedding to be loose and uncompacted. The concrete pipe standard installations also call for the outer bedding to be compacted so it can support the vertical load carried by the sidefill. With soft middle bedding and compacted outer bedding and haunch zone, a round conduit will settle further into the soft bedding as the final backfill is added over the top of the conduit and develops vertical support under the haunches. This results in a pressure distribution on the bottom of the conduit (Figure 8.4) similar to Type 1, 2, and 3 installations with the Heger pressure distribution (Figure 5.12). Good

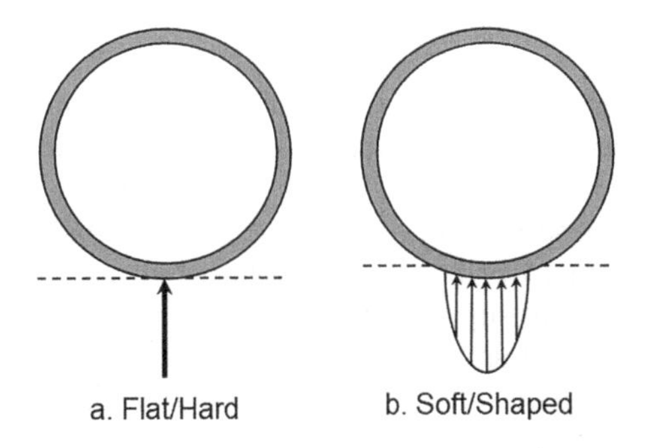

Figure 8.3 Hard versus soft bedding.

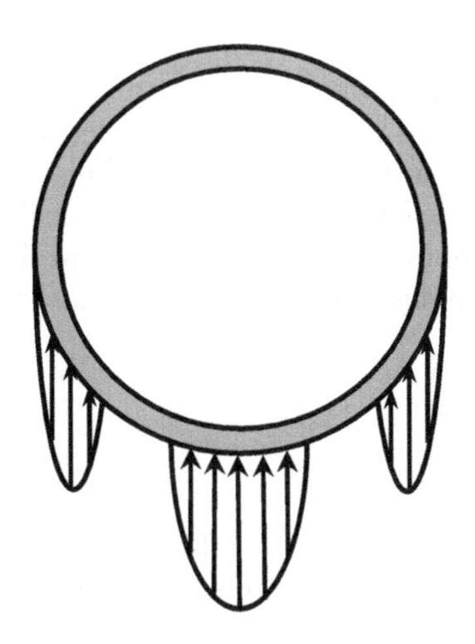

Figure 8.4 Pressure distribution with soft bedding and compacted Haunch zone.

compaction in the haunch zone is key to controlling the bending moments and cracking in a rigid pipe.

The bedding is an important layer in the process of installing circular conduits. As shown in Section 7.5.2.1, the settlement ratio is dependent primarily on the firmness of the bedding, varying from 1.0 for unyielding soil to 0.3 for yielding soil. Figure 5.4 shows that the VAF for embankment loads can be 1.7 for unyielding bedding but drops to 1.4 for yielding soil.

Bedding for vertical elliptical conduits should follow the same guidelines as for circular conduits.

8.4.2.2 Nonround conduits

Nonround conduits include three- and four-sided rectangular conduits, as well as arch and horizontal elliptical shapes.

Bedding for four-sided box conduits should be uniform and level. AASHTO Construction requires 3 in. of bedding material with a maximum particle size of 3/8 in. (10 mm) and less than 10% passing the No. 200 (0.075 mm) sieve. Compaction of the bedding should be limited to assure a compliant surface.

The footings for three-sided conduits will carry substantial vertical loads. The bedding material and compaction levels should be specified by the engineer.

Arch pipes, both concrete and metal, have large radius curved bottoms. It is critically important to provide some form of shaped bedding for these pipes to avoid hard point support at the invert. This bedding can be shaped to match the conduit radius. AASHTO Construction, Article 26, for installation of metal conduits requires shaped bedding in the form of a V under such shapes if the span exceeds 12 ft.

8.4.3 Pipe laying

After completion of the bedding, the pipe can be installed. This consists of joining the conduit sections to previously installed sections. Check that the pipe barrel is uniformly supported by the bedding and that the bells holes are adequate such that the joint does not carry vertical earth load. If the pipe is gasketed, the joint should be checked to confirm that the gasket has not been displaced. Detailed descriptions of all these tasks are outside the scope of this book.

8.5 BACKFILLING

The first two stages of backfilling after the conduit is installed are haunching and sidefilling. These are both important steps in constructing the soil-conduit system.

Embedment material must be worked into the haunch zone manually. For small-diameter conduits, this may be accomplished by placing a portion of the embedment soil and working (pushing) it into the haunch with shovels or tampers. Shovel slicing is a technique where the shovel blade is repeatedly pushed into the embedment to move it into the haunch area. Tampers with a flat face can be effective in working fine-grained soils into the haunch. The effectiveness of these techniques varies with the type of embedment soil. Class I crushed rock can be difficult to move into the haunch. Class I rounded materials move more easily and may flow into the haunch with minimal tamping. It is prudent to confirm the effectiveness of the haunch technique at the beginning of a project by excavating a short length of trench for inspection. Experience indicates that the small triangular area where the bedding, conduit, and haunching meet cannot be fully filled, but there should be no significant voids. For conduits where the sidefill below the springline is placed in more than one lift, the haunching operation should be conducted for each lift. Compaction of the sidefill below the haunch is best accomplished by compacting away from the conduit first. As the compactor moves toward the conduit, the confinement provided by the outer soil will continue to move the inner soil into the haunch zone. During the early stages of this process, procedures should be developed to avoid raising the pipe off the bedding, which is possible with lighter conduits. It may be necessary to use relatively light compaction equipment close to the conduit.

The placing and compaction of the sidefill and top fill should proceed with lift thicknesses and compaction levels as specified in the contract documents. Backfill should be brought up uniformly on both sides of the conduit. Compact the backfill for the full width of the trench or for embankment installations to the width specified. Compaction equipment should not contact the conduit. Heavy compaction equipment should not be used directly over the conduit until enough soil has been placed to prevent damaging or distorting the conduit (see Section 9.6 for a case where heavy compaction equipment caused damage.).

The final backfill imposes load on the conduit but does not impact the performance of the soil-conduit system. The soil used and compaction required near the ground surface are primarily based on the use of the surface over the conduit, be it a public way, recreational space or other open space.

8.6 CONSTRUCTION MONITORING

Monitoring conduit installations during construction is always desirable, particularly at the start of construction. The cost of repairing or replacing improperly installed conduits can greatly exceed the original installation cost.

8.6.1 During construction

Reviewing materials and procedures with the contractor should be a routine preconstruction activity. During this process, gradations for candidate foundation and embedment zone materials should be submitted to the engineer for review and approval. The party responsible for making decisions on when to remove unsuitable material from the trench bottom, the suitability of the bedding before pipe laying, and other decisions should be reviewed and agreed to. Lift thickness and compaction equipment should be identified. The use of trench boxes, if necessary, should be reviewed to ensure the backfill can be placed and compacted in accordance with the project specifications.

Close monitoring during the beginning of the installation process can prevent the repetition of poor construction practices as the project proceeds. Items to observe include excavation procedures, conduit handling, and temporary placement of excavated material away from the trench, bedding placement, and compaction, techniques for working embedment material into the haunch zone, and placing and compacting the sidefill and topfill. Deflection levels in flexible conduits should be checked after completing the first segment of a project, and appropriate action taken if the prescribed limits are not met. See Section 9.3, where failure to act on a report of overdeflection allowed inappropriate procedures to persist throughout the project.

Compaction levels in North America are typically specified as a percentage of the maximum standard Proctor density. Establishing the number of coverages by the compaction equipment can be useful for both the workers and inspectors. Once established, the engineer may set the number of passes as the approval criterion to minimize testing and give the workers certainty about their tasks.

8.6.2 After construction

It is prudent to conduct inspections after installation is completed, especially if monitoring during installation is limited. AASHTO Construction includes postconstruction requirements for inspecting concrete, metal, and thermoplastic conduits. Postconstruction inspections can consist of worker-entry and visual inspections if the conduit is large enough. Smaller conduits can utilize remote video for inspections. The first decision for postconstruction inspection is when to do it. It has been long known that soil around a conduit will settle in the period immediately after construction. This settlement is probably mostly due to rain providing water to filter down through the embedment, but it can also be caused by vehicular traffic or other events over or around the conduit. It is impractical to wait for a rainstorm, so many contracts impose a wait of 30 days after construction before conducting a final inspection. However, inspection should be completed prior to construction above the conduit, such as paving, is completed.

There are many standards for inspection and testing of conduits after installation. Standards can be general or specific to the type of conduit and the end application. Appropriate standards should be used to ensure attention to key details of conduit condition and to consider safety concerns, such as working in confined spaces for worker-entry inspections, dealing with high-pressure water or compressed air for leak testing, or other potentially dangerous activities. Common conditions to be noted during inspection are listed here, though this list should not be considered exhaustive. Standards cited in the contract documents should be referenced for more complete lists.

- Grade and alignment – Using string, laser, or visual sighting, a conduit should be checked to determine if the alignment is as specified. Gravity flow conduits should maintain a constant grade and water should be able to flow smoothly over joints. Improper grade could mean settlement has occurred or uneven longitudinal support, which could lead to beam breaks in small-diameter conduits or joint leaks.
- Joints – Many types of defects can be detected at joints:
    - Leakage – leaking joints may be identified by observation of water intrusion, but staining, silt infiltration, and displaced gaskets can also indicate leakage. Infiltration or exfiltration of water through joints can result in loss of embedment soil and long-term performance issues. Misplaced gaskets can be identified by observation or a feeler gage inserted into the joint.
    - Broken joints – bells and spigots in rigid pipe can break, which can result from a displaced gasket, poor alignment, improper bell holes, or other deficiencies.
- Excess deflection in flexible conduits and longitudinal flexural cracks in rigid conduits – Longitudinal flexural cracking and deflection beyond specified limits can be a sign of improper or lack of compaction of embedment, excessive compaction of embedment, poor haunching, heavy construction traffic during installation, or other causes. Cracking and deflection result from high bending moments. Typical limits for deflection and longitudinal crack width are 5% and 0.01 in., respectively. These deficiencies require investigation as to their causes and expert assessment to determine whether they require remediation.
- Diagonal tension cracks – Diagonal tension cracks are longitudinal cracks through the wall of a rigid conduit (see Figure 3.6) and are usually associated with excess vertical load, which can result from poor bedding or poor embedment support.
- Radial tension cracks – Radial tension cracks occur on reinforced concrete conduits when the cover concrete splits in a plane through the reinforcement (see Figure 3.7). This type of failure is rare.

- Local buckling – Local buckling can occur in corrugated thermoplastic conduits. Local buckling in the liner alone is not likely to affect the long-term performance of the conduit, but local buckling or crimping of the conduit could indicate an overload due to installation issues, such as those noted for deflection.

8.6.2.1 Worker-entry inspections

Worker-entry inspections are possible for larger conduits. Some standards limit man-entry to conduits with 30 in. (750 mm) or larger diameter or span. Inspectors must follow all appropriate laws, regulations, and guidelines for personnel safety.

Man-entry inspectors should be equipped with cameras, straight edges, and tape measures. Inspection notes should be recorded on forms prepared beforehand, either on paper or electronically. A form to be filled out for each section of conduit showing deficiencies is recommended. Conduit sections without deficiencies should be noted to confirm that each length has been inspected. Notes should include the location of deficiencies within a section of conduit. A clock position is typically used to identify the circumferential position, but a system for identifying the 3 o'clock and 9 o'clock locations must be established (for example, in a culvert, the 3 o'clock position may be identified as on the right when looking downstream). Lengths are often numbered to identify the longitudinal position.

8.6.2.2 Video inspection

Video equipment for inspecting conduits has become very sophisticated. The simplest video inspection consists of a camera looking straight down the conduit. This is effective for observing many conditions, but the inability to orient the camera to look directly at a defect is limiting, and in some cases, the camera is not able to locate a defect. Depending on the lighting, it can be difficult to perceive the deflections with a simple straight video camera. More advanced video equipment which can look in all directions inside the conduit is preferred. Some video inspection units can measure conduit deflections as well.

8.6.2.3 Deflection monitoring

Deflection monitoring can be as simple as pulling a mandrel that has been sized to not pass through an overdeflected conduit. This provides minimal information on the general nature of the installation. Overdeflected areas will stop the mandrel and may need to be repaired to allow the mandrel to pass through to test the rest of the length of conduit. Actual measurement of the pipe shape is the preferred method to collect deflection data. Data collected can be the vertical diameter, both vertical and horizontal diameters, or the full conduit shape.

Taking diameter measurements at fixed intervals along the length of a conduit is useful in evaluating the quality and consistency of the embedment support to the conduit.

8.6.3 Inspection follow up

The inspection agency should submit a report with the findings of the inspection to the owner and or engineer. The contract documents should include the process for determining remedies for any deficiencies in the report.

REFERENCES

AASHTO (2017) *AASHTO LRFD Bridge Construction Specifications*, 4th Edition. AASHTO, Washington, DC.

Howard, A. (2015) *Pipeline Installation 2.0*, Relativity Publishing, Lakewood, CO.

Chapter 9

Experiences with conduit performance

Engineering often progresses by looking back at projects that developed problems and evaluating how these problems could be avoided in the future. This learning process functions on a large scale, where lessons learned can lead to changes in engineering, materials, or construction practices, or on a small scale, where an engineer, manufacturer, or contractor learns a lesson. Projects consist of the following:

- Design – The engineering that determines the materials required to complete a project and the manner in which they should be assembled.
- Materials – The products specified in the engineering phase.
- Construction – The implementation phase where the supplied materials are incorporated into the project in accordance with the project specifications and the manufacturer's instructions.

Sometimes responsibilities for a project overlap. For example, a manufacturer may provide a proprietary design for a portion of a project, or construction may involve supervision by the engineer or a third party. In these latter cases, both parties must understand their own role and protect against leaving a gap.

This chapter presents several projects where the authors were directly involved in investigating what went wrong and how responsibility might be assigned. Minimal details are provided for each project to keep the reader focused on the primary cause(s) of the problems. Some have been written up in the literature while others have never been previously published. Project names and personnel are not identified to preserve privacy. It is not as important to know who did it as to see how the problem can be avoided in the future. The order in which the cases are present is random and not meant to suggest one has a more important lesson than another.

DOI: 10.1201/9780429162619-9

9.1 UNDERSTANDING SOIL-CONDUIT INTERACTION (UNPUBLISHED)

9.1.1 Background

Stormwater retention chambers have come into use as environmental regulations require that stormwater drainage be allowed to seep into the ground rather than be carried off-site through drainage systems. This mitigates the impact of the vast areas that are covered with pavement or buildings. A major focus of stormwater retention is storage volume – a place to hold stormwater until it has time to seep into the ground. While traditional conduits – concrete, metal, and plastic pipes and chambers serve this purpose, open-bottom arches offer storage volume and a large surface area for stored water to seep back into the ground. A typical installation of a thermoplastic arch chamber is shown in Figure 9.1. The stone backfill used between and over the chambers is quite stiff, i.e., the modulus of elasticity of the stone (M_s) is high. The low cross-sectional area and low long-term modulus of elasticity of the conduit result in a much lower stiffness than the backfill. This differential stiffness results in a load path through the stone, much like the load path discussed for thermoplastic pipe under live loads (Figure 7.6).

Finite element modeling demonstrates that the same load path controls the path of the live load. In Figure 9.2, the light areas indicate the flow of the load through the soil and around the chamber.

The economics of stormwater retention systems are based on the fraction of a system that is available for storing water. A manufacturer sought to improve storage volume by bridging the space between the polyethylene arches and thus increasing storage volume (Figure 9.3). After development and testing, the product was widely distributed and installed; however, the chamber beds began collapsing within a few months after installation. Since the chamber beds were large, the failures manifested themselves by causing a collapse of the entire ground surface over a bed, sometimes as large as 10,000 ft^2 (930 m^2).

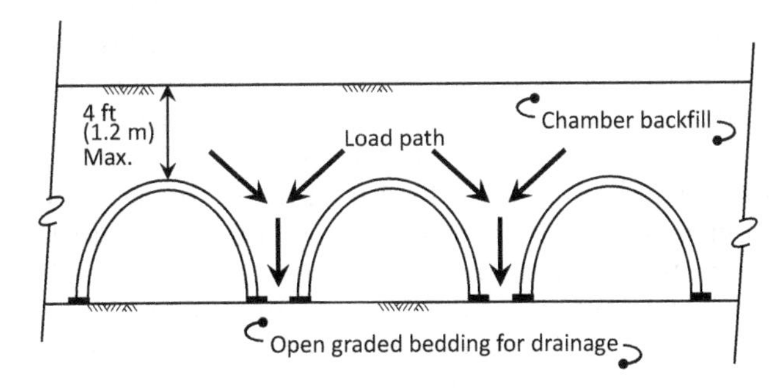

Figure 9.1 Typical stormwater chamber installation.

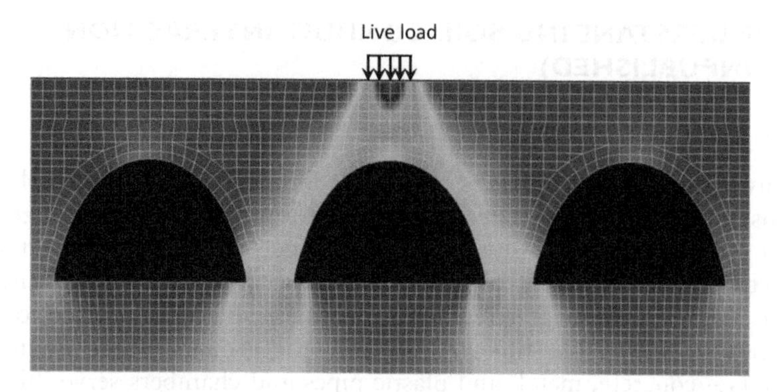

Figure 9.2 Live load path around thermoplastic stormwater chamber.

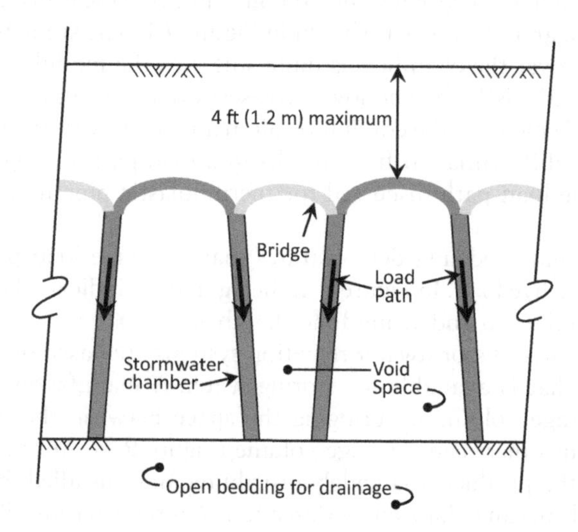

Figure 9.3 Stormwater retention system with increased storage volume.

9.1.2 Investigation

Investigation and calculations showed flaws in the design concept and in the validation testing conducted by the manufacturer. As discussed in Chapters 3 and 7 and above, thermoplastics will creep under constant load as the apparent modulus decreases over time. In the typical installation, Figure 9.1, the primary load path is through the chamber backfill, as this is the stiffest element in the chamber bed. This load path limits stresses in the chamber, and the chamber stresses continue to decrease over time, just like thermoplastic pipes. In the installation with the bridge between the chambers, all load must travel through the chamber legs to the bedding layer. In this scenario, the stresses in the chamber are high and do not decrease with time.

In the product development phase, a judgment was made that the live load condition would be the controlling load for the chambers, which is typical for shallow buried conduits. This judgment led to testing chamber performance under short-term vehicle loads. Live loads can be a limiting condition for chambers in Figures 9.1 and 9.3; however, earth load can also control the design. This is especially true for the chambers with the bridge. Under the higher stresses in the legs, the chambers underwent creep deformations over time and eventually buckled and collapsed; however, the short-term live load testing did not evaluate this possibility.

9.1.3 Conclusions

The collapse of the stormwater chambers in Figure 9.3 occurred due to an inadequate understanding of soil-conduit systems and the time-dependent nature of thermoplastic properties. Figure 5.3 demonstrated how the VAF on corrugated PE conduits is substantially less than 1.0. The *VAF* is initially low and reduces further with time under load. The same is true for the thermoplastic stormwater chambers in Figure 9.1, where the embedment carries the majority of the load, and the proportion increases over time due to creep in the plastic. Eliminating the backfill between chambers increased the load on the chambers significantly. The *VAF* for Figure 9.3 is the chamber spacing divided by the chamber width -substantially greater than 1.0. Creep over time led to compression failure or general buckling.

The failure might have been detected during validation testing if tests had been carried out to evaluate the chamber performance under long-term earth load conditions. In this case, the assumption that the governing load condition was a short-term live load led to the design of a test program that would not actually test the capacity of the system.

9.2 FAULTY CONSTRUCTION PRACTICES, INCOMPATIBLE EMBEDMENT MATERIALS (SELIG AND MCGRATH, 1994)

9.2.1 Background

A new sanitary sewer system consisted of 7 miles (11 km) of 20 to 42 in. (500 to 1,100 mm) diameter flexible fiberglass (also known as glass reinforced plastic, or GRP) pipe. The sewer lines were installed mostly below the water table. Native soil was a uniform fine sand. Embedment around the pipe was a uniform crushed stone to a height of 70% of the pipe diameter. Sand fill similar to the native sand was used over the stone. Changing backfill partway up a conduit is sometimes called a split installation.

After construction, failures occurred that generally consisted of the tops of the pipes collapsing into the pipe. Due to the high water table, water would rush into the opening created in the pipes and carry off large quantities of overburden, eventually resulting in sinkholes at the surface, sometimes large enough for several vehicles to fall into.

9.2.2 Investigation

The investigation consisted of field trips to observe the conditions at failure sites and during excavations at locations without failures.

9.2.2.1 Dewatering/trench sheeting

The trenches for pipe installation were dewatered through well points outside the trench. The trench walls were supported with steel sheet piling to provide a safe workspace. However, observations and interviews showed that dewatering was ineffective. In addition to the well points outside the trench, a dewatering pipe, working on the same vacuum system as the well points, was left in the trench to remove the water flowing in, but this pipe often drew in air, which compromised the vacuum needed to successfully operate the well point system. As a result, the groundwater level was not lowered below the bottom of the trench, and groundwater was present behind the sheeting. The sheet piles had simple lap joints that separated laterally as they were driven into the ground, resulting in open joints. Groundwater flowed into the trench through the joints, carrying native sand with it and leaving voids behind the sheeting. When the sheeting was removed after pipe installation, the embedment moved into the voids behind the sheeting, causing a loss of support to the pipe and increased deflection.

9.2.2.2 Migration of sand into voids in stone

Investigation revealed that the sand backfill placed over the stone had washed down into the top layer of the stone embedment. This should have been anticipated, as the gradations were incompatible, as shown in Figure 9.4. According to practice guidelines, the embedment material gradation should have fallen into the shaded area to prevent migration of the native sand. With the crushed stone backfill, a filter fabric should have been provided to maintain separation.

The incompatible gradations resulted in migration of the sand trench backfill into the stone embedment, which resulted in a loss of support for the trench backfill; however, since migration could not occur directly over the pipe, vertical earth load became concentrated over the crown of the pipe, resulting in excess moments and deformation (Figure 9.5). This led to the collapse of the top of the pipe.

9.2.3 Conclusions

The poor dewatering techniques resulted in a loss of support to the sides of the pipe and high deflections. The migration of sand into the top of the stone embedment concentrated the earth load on the top of the pipe. Steps that should have prevented these issues include:

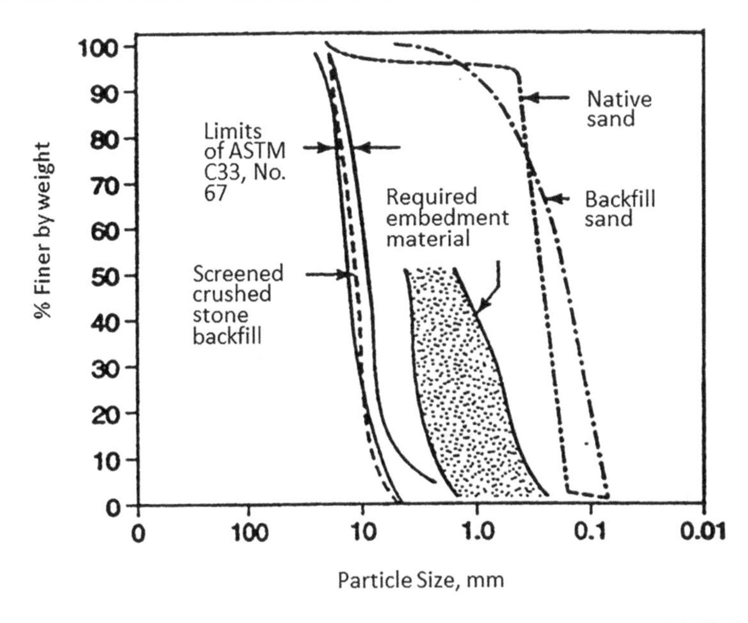

Figure 9.4 Relative gradation of embedment stone and sand backfill (Selig and McGrath, 1994)

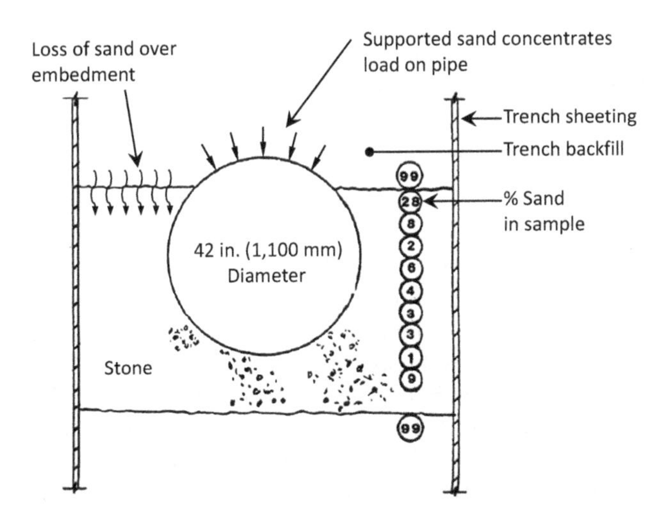

Figure 9.5 Migration of sand backfill into stone embedment.

- Observations during construction should have noted the water flowing into the trenches through the sheet piling joints. Improved dewatering techniques should have been used to lower the groundwater below the trench bottom. Further, the use of sheeting with locking joints could have avoided the gaps through which sand flowed into the trench.

- An alternate backfill gradation or a filter fabric could have been used to separate the open-graded stone from the sand materials surrounding it.
- Extending the stone embedment material to at least 6 in. (150 mm) over the top of the pipe would have minimized or eliminated the concentration of load that would inevitably occur when sand migrated into the stone.
- Installation of short lengths of impermeable pipe zone embedment at locations along the pipelines could have reduced the flow of water through the embedment parallel to the pipeline. Reduced flow would have in turn reduced the risk of fines migrating into the embedment.

9.3 POOR CONSTRUCTION MANAGEMENT, POOR CONSTRUCTION PRACTICES, FAILURE TO SEEK GUIDANCE (MCGRATH AND SELIG, 1994)

9.3.1 Background

A nuclear power plant design called for four 12 ft (3.7 m) diameter pipelines to carry cooling water, each line being approximately 600 ft (184 m) long. Burial depths generally varied from 5.5 to 20 ft (1.5 to 6.1 m) but were mostly less than 10 ft. At the end of construction, the pipes were severely deformed, and the joints had delaminated.

The project was a complex combination of many site utilities and multiple contractors. Construction management was provided by a site organization that monitored construction and specification compliance.

9.3.2 Specifications

The pipe selected for the project was filament wound fiberglass (GRP) with a wall thickness of 1.1 in. (28 mm), except at the joints where the bells were thickened to approximately 5 in. to provide a seal over the O-ring gaskets. The pipe stiffness of the barrel was low, $PS_{B\text{-}US}$ = 5 psi and $PS_{B\text{-}I}$ = 0.64 kPa, and the bell stiffness was high, $PS_{B\text{-}US}$ = 400 psi and $PS_{B\text{-}I}$ = 51 kPa.

Backfill was a poorly graded sand (SP per ASTM D2487) with specified compaction to 85% relative density. Compaction within 6–18 in. (150–450 mm) of the pipe was to be accomplished with hand tampers. The site organization believed that meeting the density specification was required to prevent liquefaction in the event of an earthquake.

Project documents required monitoring deflection during backfilling. If the horizontal diameter decreased by more than 3%, the required density was to be reduced to 75% relative density, and the site organization was to be notified.

Project drawings showed many site utilities and pipelines crossing under and over the circulating water lines. One line was shown to be essentially in contact with the top of the circulating water lines.

9.3.3 Investigation

The primary compaction equipment for the pipe embedment was a double drum vibratory roller applying peak vertical dynamic compaction stresses of 37 psi (255 kPa). Field reports showed that the 3% limit on inward deflection was exceeded on the very first pipe length installed. Despite the specification instruction to reduce the compaction requirement to 75% in this situation, the site organization told the contractor to continue to meet the original compaction specification without providing guidance on how to achieve the required density without exceeding the deflection limit, such as using lighter compaction equipment near the pipe.

Loud "shotgun-like" noises were heard during backfill compaction. Upon discussion, the site organization decided the noises must be typical of fiberglass pipe. The manufacturer was not consulted on this decision.

A crane weighing 390,000 lbs (1.7 MN) was repeatedly moved across the pipelines. A crane weighing 2,000,000 lbs (8.9 MN) was moved across the pipelines at least once.

At the end of construction, 80% of the pipe lengths exceeded the 3% specification limit on horizontal inward deflection. Many of the pipe sections were cracked or crazed, either from compaction forces or inadequate protection from live load forces during the crane crossings. The pipes were often distorted, being pushed in from one side far more than the other or being squeezed inward from both sides in the shoulder area of the pipe, taking a shape like a pear. Many pipe joints were delaminated. The manufacturer reported that almost identical pipes had been successfully installed on a similar project.

Laboratory and field density data are presented in Figure 9.6. All three tests, i.e., minimum, field, and maximum, showed predictable variation. Standard deviations for the three conditions were each about 2 pcf (0.30 kN/m^3). This is significant in respect to relative density, as 2 pcf (0.3 kN/m^3) represents about 11% relative density, and the consequences are that many tests that should have represented good compaction showed less than 85% RD and were required to receive more compactive effort resulting in further deformation of the pipes.

Field reports indicated that the double drum vibratory compactor was operated within 6 in. (150 mm) of the pipes rather than the 24 in. (600 mm) required by the specifications. Figure 9.7 suggests that the pressure applied with the compactor 6 in. (150 mm) from the pipe was about seven times the pressure if the compactor was held 24 in. (600 mm) from the pipe (0.2 versus 0.03 of the applied pressure in the figure). Computer modeling showed that pipe response would have been acceptable had the specification limit been followed and that upward deflections comparable to those observed in the field would be expected for the 6 in. (150 mm) condition. The use of lighter equipment near the pipeline likely would have achieved the desired density without excessive distortion. It is also likely that accepting a slightly

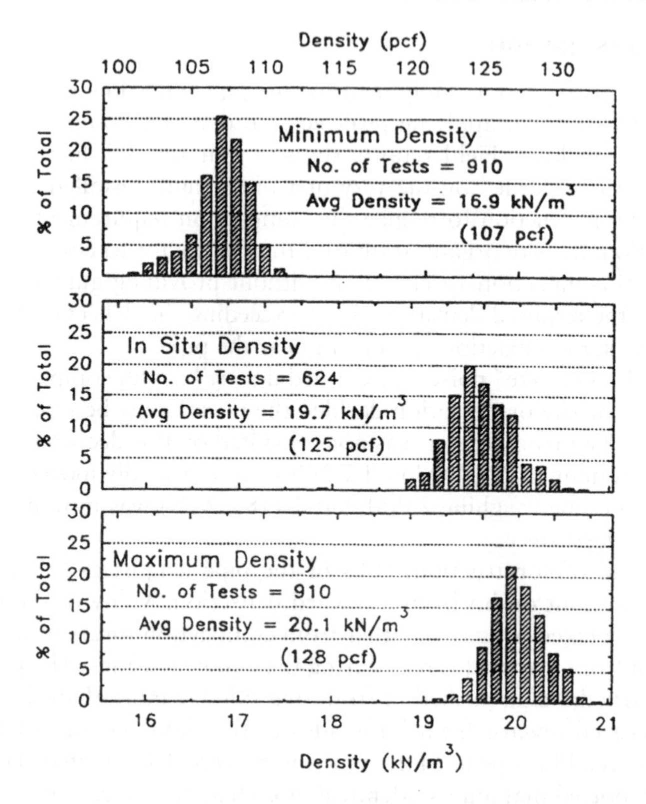

Figure 9.6 Field and laboratory density test results (McGrath and Selig, 1994)

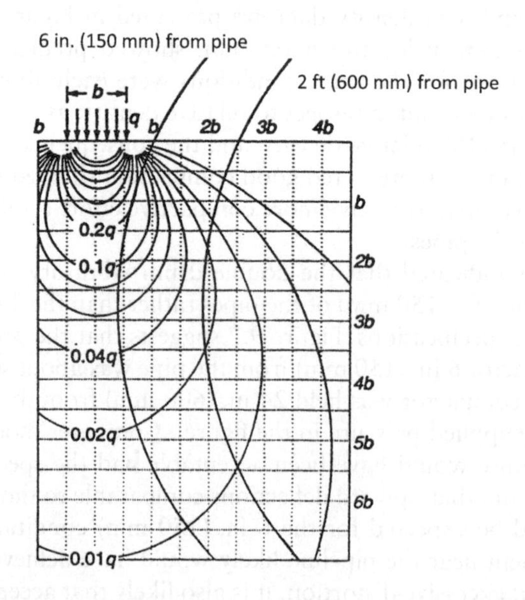

Figure 9.7 Pressure exerted on pipe by vibratory roller (McGrath and Selig, 1994)

lower density in the vicinity of the pipeline, as permitted in the specifications, would not have compromised the overall resistance to liquefaction.

9.3.4 Conclusions

The primary responsibility lies with the construction management, which had reports of excess deformation right from the start of construction and failed to take action to address the issues. Procedures could have been developed to correct the deficiencies. The poor construction management was exacerbated by the use of the vibratory compactor close to the pipe. Manufacturers, designers, construction monitors, and installers could all share responsibility for specific issues.

- The upward deflection and distortion of the pipes in the circulating water lines resulted from excessive compaction forces. The upward deflection was noted and reported to the site organization early in the project but continued throughout. The site organization was focused on the density requirement, and it did not consider alternate compaction options and did not consult the manufacturer for advice. There is no indication that alternate compaction methods or density criteria were ever considered for the area close to the pipes.
- The pipes were not protected from live load effects during crane crossings. Placing temporary fill over the pipes to increase the depth of laying stiff mats over the pipe to spread the load over a wider area could have eliminated the risk to the pipes.
- The site organization was unfamiliar with fiberglass pipe but never consulted the pipe manufacturer for advice during construction.
- Field reports and computer modeling suggest that the compactor was operated too close to the pipe, contributing to the excessive deformations.
- Also related to excessive compaction stresses, the "rifle shot cracks" heard during backfilling were the thick bell joints cracking and delaminating. The issue was discussed on-site, but the manufacturer was not consulted to investigate the incorrect conclusion that it was typical of fiberglass pipe.
- The decision to use a relative density criterion for evaluating backfill compaction rather than a Proctor-type criterion was inappropriate and likely led to excess compactive effort.
- The damage at the joints was likely exacerbated by the significant stiffness variation between the pipe barrel and the bells. The barrels deflected significantly more than the bells, resulting in high longitudinal stresses. This stiffness differential may not have caused a problem if proper construction controls to limit pipe deflection had been used but should generally be avoided.

9.4 CONTRACTOR NOT FOLLOWING INSTALLATION INSTRUCTIONS (NOT PUBLISHED)

9.4.1 Background

An engineer was asked for a special concrete pipe design for an installation with deep fill. The engineer submitted a design requiring stirrup reinforcement to address the high shear stresses in the pipe. The pipe manufacturer asked if a design without stirrup reinforcement would be feasible, to which the engineer asked if the manufacturer could ensure stringent installation instructions could be followed. With the agreement to enforce such instructions, a new design was provided and submitted. The design called for high-quality embedment zone backfill carefully worked into the haunches and compacted. The manufacturer sent a staff engineer to the project to establish the backfill procedures in accordance with the agreed-upon specifications. Once the construction was underway without problems, the staff engineer left the project. Later, the contractor called the manufacturer to report that the concrete pipes were failing.

9.4.2 Investigation

Trips to the site found that the pipe in the portion of the project installed under the direction of the manufacturer's engineer was performing as designed. The pipe installed after the engineer's departure had wide flexural cracks and diagonal tension cracks. When cores were drilled into the pipe, they fell into the haunch zone where there was no backfill. At this point, the contractor admitted they had stopped using the specified high-quality backfill and instead placed large shot-rock from blasting operations as pipe embedment. The cores indicated that none of the shot-rock had flowed into or been worked into the haunch zone.

9.4.3 Conclusions

The contractor attempted to save money by substituting the shot-rock for the specified embedment material and was totally responsible for the problems. However, in the interest of reaching a quick settlement without litigation,

- the contractor took on the expense of removing the damaged pipe sections, which required removing all embedment and backfill material over and around the damaged portion of the pipeline;
- the pipe manufacturer supplied new pipe with stirrups to minimize the risk of future problems; and
- the design engineer did not seek reimbursement for the travel and consulting costs incurred in investigating the failure and designing the replacement pipe.

In summary, all parties, even those without liability, incurred costs and had to ask themselves if the attempt to reduce the cost of the project was worth it.

9.5 REMEDIATION OF SOFT IN SITU SOILS (HEGER AND SELIG, 1994)

9.5.1 Background

A new highway embankment required routing a small stream through a concrete culvert. The stream bed was soft and wet and not suitable as a foundation for the proposed culvert. The contract documents called for replacing the soft subgrade only directly under the culvert pipe. After installation, the pipe developed cracks that substantially exceeded the service load limit of 0.01 in.

9.5.2 Investigation

The embedment design for the pipe installation is presented in Figure 9.8. The layer of soft in situ material, which overlaid rock, has been highlighted. This figure shows the following:

- The in situ material was only removed sufficiently to make room for pipe installation. The pipe was then laid on sand bedding, which in part was supported on the soft in situ material.
- The compacted granular embedment was then placed on top of the soft in situ material, and the embankment was constructed around and over the embedment.
- Computer modeling of the installation indicated that the soft in situ material would compress, causing a loss of support to the pipe embedment zone and the adjacent embankment material. Settlement of the material beside the conduit caused negative arching, substantially increasing the load on the conduit. The calculated *VAF* for the installation was about 1.7, while a typical concrete pipe installation has a *VAF* of about 1.4. Further, the pipe was only firmly supported in a relatively narrow width at the invert. The haunches were essentially unsupported due to compression of the soft material. The indirect design method would also have identified this design condition. The settlement ratio (Table 7.2) for unyielding bedding would be 1.0, which means the *VAF* would be 1.7, confirming the computer model. A *VAF* of 1.7 would cause the observed failure.
- The bedding thickness could have been increased to mitigate the concentrating effect of the underlying rock.

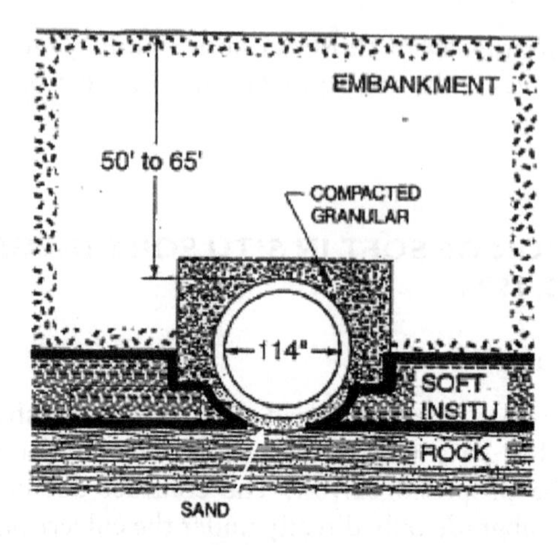

Figure 9.8 Design drawing for pipe bedding and embedment (Adapted from Heger and Selig 1994 Figure 1, p. 47. Copyright National Academy of Sciences. Reproduced with permission of the Transportation Research Board)

Computer modeling confirmed that these conditions resulted in a failure of the pipe.

9.5.3 Conclusion

The design drawing, as shown in Figure 9.8, is not suitable for actual construction from both structural system and constructability viewpoints. This detail suggests a lack of familiarity with conduit design, soil-conduit systems, and construction practices.

- Leaving soft in situ material under the embedment immediately adjacent to the pipe allowed settlement, which altered the structural system and imposed a significantly greater load on the conduit.
- The pipe is firmly supported on a layer of sand bedding resting on rock, but the sand bedding is only supported on rock over a relatively narrow area near the invert and then supported on the soft in situ material. This resulted in a hard narrow bedding condition.
- It would be difficult and time-consuming to remove such a limited portion of the soft material to such an exact shape. The designer should have addressed this.

If the project were constructed in this fashion, the pipe would be overloaded, and failure would be expected. It is likely that the as-built profile varied from that shown in Figure 9.8, but specific field details were not available. The soft in situ material should have been removed beyond the full diameter of the pipe, the pipe embedment, and the lower sidefill

(Figure 7.22) to create a support that would limit the arching of earth load onto the pipe. The AASHTO LRFD Construction Specifications (AASHTO, 2017) specify minimum compaction requirements for the lower sidefill to avoid this scenario.

9.6 INADEQUATE SPACE FOR COMPACTION AND INADEQUATE COVER DEPTH DURING COMPACTION (UNPUBLISHED)

9.6.1 Background

A number of trenches were used to accommodate multiple reinforced concrete and vitrified clay sewer conduits during fast-track construction of a new hospital. The sewers were installed, and construction of the reinforced concrete building proceeded overhead. Subsequent testing of the conduits indicated that a significant number were badly damaged, with none of the water introduced upstream making it to the downstream end of some conduits. There was significant leakage from conduits at all depths of cover. Replacement of the conduits was expensive given that work had proceeded overhead.

9.6.2 Investigation

Calculations showed that maximum burial depths exceeded the limits based on the code in place when the pipes were designed; however, this inadequacy did not explain the severity of the damage and the problems experienced with some of the shallowly buried conduits.

The investigation showed that the construction drawings were unclear and did not adequately define the minimum distances required between the conduits to allow proper placement and compaction of backfill. The trench width during construction was inadequate to accommodate the multiple conduits with adequate spacing. The construction specifications demanded a high level of compaction of the backfill near the top of the trench (at the ground surface) – even when this was just above some of the shallow buried conduits. No limits were defined regarding the type of construction equipment that the pipelaying subcontractor was permitted to use to achieve the compaction. Finally, photographs taken during exhumation and replacement of the damaged conduits showed that adjacent conduits were almost touching in places and that there was no backfill between adjacent, parallel sewers.

9.6.3 Conclusions

Damage occurred because of inadequate soil support between closely spaced conduits and subsequent failure to achieve the soil support required by design. The absence of sidefill in some locations led to concentrated loading, close to a D-loading and circumferential bending moments much greater than design values. Segments of conduits at shallow cover were also likely damaged because of the compaction equipment used to achieve the high densities required by the specifications.

The design failed to provide adequate guidance on minimum spacing for adjacent pipes and control of compaction equipment immediately above the conduits. Oversight by the engineer, prime contractor, or subcontractor for conduit installation should have identified the problems during construction, providing an opportunity to correct the situation prior to continuing construction over the conduits.

9.7 INADEQUATE UNDERSTANDING OF BUCKLING CAPACITY (MOORE ET AL., 1995a, 1995b)

9.7.1 Background

A series of long-span corrugated steel culverts were used for county roads in the late 1960s and 1970s. Some of these were constructed at shallow cover – either over the entire roadway or at the edges of the sloping embankment. One culvert collapsed during structural rehabilitation. Inspections revealed that some of the remaining structures had wavy deformation patterns across the top of the structure. The design of these structures, according to the accepted standard, did not consider the global buckling strength limit because it was believed that special features like thrust beams and circumferential ribs prevent buckling.

9.7.2 Investigation

The structures were designed and constructed before the controlling bridge design code addressed their potential for buckling. The excessively conservative buckling theory at the time (a beam-on-elastic-spring model) would have predicted that the expected thrust exceeded the buckling thrust. Continuum buckling theory was used to review the stability of several culverts in the county, to check the adequacy of their resistance to global buckling. Calculations showed that the shallow cover conditions in some culverts reduced the critical thrust to values near or below the expected levels of thrust at these locations. The buckle wavelengths predicted by continuum buckling theory were similar to those observed in the field.

9.7.3 Conclusions

A number of long-span corrugated steel culverts were vulnerable to global buckling, a failure mode that was not considered during their initial design. Continuum buckling theory identified such culverts and allowed selective remediation. Structural strengthening of one structure involved addition of further cover soil over the outer ends of the culvert, where the previous sloping embankment geometry was calculated to have produced shallow cover conditions and inadequate soil support.

REFERENCES

AASHTO (2017) *AASHTO LRFD Bridge Construction Specifications*, 4th Edition, AASHTO, Wahington, DC.

Heger, F.J., and Selig, E.T. (1994) *Rigid Pipe Distress in High Embankments Over Soft Soil Strata, TRR 1431*, Transportation Research Board, Washington, DC.

McGrath, T.J., and Selig, E.T. (1994) *Backfill Placement Methods Lead to Flexible Pipe Distortion, TRR 1431*, Transportation Research Board, Washington, DC.

Moore, R.G., Bedell, P.R., and Moore, I.D. (1995a) Investigation and Assessment of Long-Span Corrugated Steel Plate Culverts, *Journal of Performance of Constructed Facilities*, Vol. 9, No. 2, May, pp. 85–102, ASCE, Reston, VA.

Moore, R.G., Bedell, P.R., and Moore, I.D. (1995b) Design and Implementation of Repairs to Corrugated Steel Plate Culverts, *Journal of Performance of Constructed Facilities*, Vol. 9, No. 2, May, pp. 103–116, ASCE, Reston, VA.

Selig, E.T., and McGrath, T.J. (1994) *Pipe Failure Caused by Improper Groundwater Control, TRR 1431*, Transportation Research Board, Wahington, DC.

Index

Pages in *italics* refer to figures and pages in **bold** refer to tables.